STRICTLY MOBILE

Strictly Mobile

HOW THE LARGEST MAN-MADE PLATFORM IN HISTORY IS CHANGING OUR WORLD

As told by twelve mobile thought leaders:

Gary Clayton
John Couch
Jennifer Haroon
Aditya Khurjekar
Manish Kothari
Ezra Kucharz
Bill Mark
Bertrand Nepveu
Bob Richards
Paola Santana
Eric Topol

and
Kevin Talbot

COPYRIGHT © 2016 KEVIN TALBOT
All rights reserved.

STRICTLY MOBILE
How the Largest Man-Made Platform in History Is Changing Our World

Edited by Kevin Talbot
Gary Clayton
John Couch
Jennifer Haroon
Aditya Khurjekar
Manish Kothari
Ezra Kucharz
Bill Mark
Bertrand Nepveu
Bob Richards
Paola Santana
Eric Topol

Illustrations by Zsofi Lang
Cover Design by Wes Montgomery
Title Artwork by Alex Parker

ISBN 978-1-61961-429-1 *Paperback*
      978-1-61961-430-7 *Ebook*

For HB, who taught me I could do anything I put my mind to.
To P, K, and D for your patience, support, and love. You are my heroes.
—KT

CONTENTS

Acknowledgments 10

Preface 12

Introduction 14

The Largest Man-Made Platform in History 16
Kevin Talbot

Artificial Intelligence: The Brain Behind the Mobile Revolution 24
Bill Mark

The Creative Destruction of Medicine 36
Eric Topol

Education's Digital Future 52
John Couch

Transforming Mobility with Self-Driving Cars 62
Jennifer Haroon

Robots and Manipulation: The Next Frontier in Mobility *72*
Manish Kothari

The Case for Autonomous Transportation *82*
Paola Santana

Pioneering the Lunar Frontier *94*
Bob Richards

The Mobile Front Lines *102*
Ezra Kucharz

How Mobile Has Created a New Way to Pay *110*
Aditya Khurjekar

Virtual Reality: The Final Platform *120*
Bertrand Nepveu

Will Technology Love Me Back? *132*
Gary Clayton

"The best way to predict the future is to invent it."

—ALAN KAY

ACKNOWLEDGMENTS

I am truly grateful to all those who have assisted with this book. I would like to express my sincere appreciation and thanks for their generous support and help.

I would like to thank Zach Obront, our publisher and the team at Lioncrest. As the scope of this project evolved they never lost interest or patience with us. We are especially grateful to Lauren Holstein who helped each of us bring our words to the pages of this book so eloquently.

The image on the cover of this book also tells an important story about how technology has transformed our lives, over a relatively short period of time, and a huge thank you goes to our illustrator Wes Montgomery who so accurately depicted 21st century technology in a 1950s living room. Thank you also to Alex Parker who collaborated and project managed the cover art. And finally, our thanks to Zsofi Lang, who captured each of us in the illustrations that accompany each chapter.

To my partners at Relay Ventures, I would like to thank you for believing in this project along with many of my other adventures including the Strictly Mobile conference that lent its name to this book. While these activities take me away from our primary business of early-stage investing, your continued support of these efforts benefit our portfolio companies and make us all better investors.

This book, however, would never have gotten off the ground without the trust that my eleven co-authors placed in me. Gary, Bill, Paola, Eric, Bob, Jennifer, Manish, Bert, Ezra, Aditya and John have lent their names and intellectual property to this book and I am forever grateful for their participation. They are all thought leaders in their fields and I am honored be included with them in this collection of essays.

And, finally I would like to acknowledge and thank my family for their patience and support. This book brings to life the future that I am so passionate about. And, now you know.

PREFACE

In 2008 when we founded Relay Ventures, an early stage venture capital fund focused exclusively on emerging mobile software, services, content, and technology, it was not a foregone conclusion that mobile would become a dominant global phenomenon. In fact, many people, with arguably more experience in the venture industry than us, thought we were making a big mistake. They told us mobile would be a niche. We thought differently. We had no idea it would evolve the way it did, but we had an inkling that it could be huge. So we struck out and labeled ourselves with the tag line "Strictly Mobile" so that no one would be confused about our focus.

Around the same time, we began to gather thought leaders together for an annual "un-conference" that we later branded Strictly Mobile. Now in it seventh year, Strictly Mobile is an invitation-only event attended by over three hundred people annually. Over the years we have delved into such topics as the digital classroom with Sal Khan, founder of the Khan Academy, mobile security with the world's most famous hacker, Kevin Mitnick, and digital health care with world-renowned cardiologist and author Dr. Eric Topol. Executives from Silicon Valley giants like Google, Apple, Yahoo, and eBay have debated on our stage. And many entrepreneurs building the next generation of start-ups have joined us to talk about their dreams and vision for the future.

As a peripheral actor in the mobile ecosystem, I am inspired by these thought leaders and entrepreneurs. They are the people creating today what we will talk about tomorrow. So, when I approached my coauthors about contributing to this project, I got a resounding positive response. The result, after much work, is a compilation of essays on a wide variety of topics all related to mobile, including self-driving cars, drones, robots, virtual reality, mobile payments, digital education, consumer health care, digital content, space exploration, artificial intelligence, and our relationship with technology.

This book would not have been possible without the willing cooperation of my coauthors who have lent their names and ideas to this project. I would like to thank Gary Clayton, Eric Topol, John Couch, Jennifer Haroon, Manish Kothari, Paola Santana, Bob Richards, Ezra Kucharz, Bertrand Nepveu, Aditya Khurjekar, and Bill Mark. I am honored to be associated with you.

It is my impression that not one person in this book works in technology solely for financial gain. Their paths all started with a vision too consuming to ignore and have led them to journeys that hold deep personal value. Whether their ambitions were fueled by a love for family, a desire to change the world, or a thirst for knowledge, I believe you will find their stories profoundly inspiring.

—KEVIN TALBOT

INTRODUCTION

It is said that a picture is worth a thousand words. There are three photographs that I like to use in my presentations to describe the transformational power of mobile on our society, our world.

The first is a contrasting photograph taken in St. Peter's Square from behind the crowd of thousands in attendance. The top half was taken in 2005, at the inauguration of Pope Benedict, and the bottom half was taken in 2013, at the inauguration of Pope Francis. As you can imagine, in the top picture from 2008 there is only one phone screen visible. And it's a flip phone. But the bottom half is a dramatically different story. It appears that only a handful of people don't have a phone, with a glowing blue screen, in their hands. In 2008, there were approximately 2 billion mobile phones in the world. By the time the bottom photo was taken there were 6.9 billion. This contrasting image is symbolic of the consumerization of computer technology in the form of the now ubiquitous smartphone—the super computer in everyone's pocket.

When I show people these contrasting and contradictory images, they immediately jump to the conclusion that the opportunity in mobile, for further innovation and investment, is over. With more mobile devices on the planet than people, do these images show us how far we have come or how far we still have to go?

Which brings me to the second photo. In it, a group of people are posing for a selfie with Pope Francis at the Vatican. Who could have ever imagined taking a selfie with a Pope? The transformation of how we communicate, the democratization of media, and the creation and consumption of content have all been made possible by mobile technologies that are always on, always connected, and always with us.

And finally, I like to finish my presentations with a photograph taken in China showing a divided pedestrian sidewalk, the right side for people using their smartphones while walking and the left side where smartphones are prohibited. This is the photograph that brings us full circle, showing us the absurdity and apparent contradictions of how technology can be taken too far. It is an example of what can be lost when we become absorbed and obsessed with the frivolous, suffering from "fear of missing out."

The march of technology and innovation has transformed industries and institutions throughout history. This is not new. What is new is the pace of change of this transformation. It will continue and it will accelerate. It will also expand and reinvent just about every industry we can imagine. It is fun and exciting. But it can also stress the fabric of what it means to live our lives in the present. This is best summed up by a paper sign recently seen stapled to a hydro pole that reads "Be present for our kids. Put down your phone. Stop checking email. Limit screen time. Turn off your TV. Read books. Go outside. Enjoy."

THE LARGEST MAN-MADE PLATFORM IN HISTORY

Kevin Talbot

KEVIN TALBOT is Co-founder and Managing Partner of Relay Ventures, an early stage venture fund exclusively focused on mobile software, services, content, and technology. Kevin has been an entrepreneur, operator, and investor in both Canada and the United States for over twenty years. From 1997 until 2010, Kevin was with the venture capital arm of Royal Bank of Canada where he was Vice President of Venture Capital and Managing Director of RBC Venture Partners. He was focused on investments in financial services, SaaS, and enterprise software throughout North America. He cofounded Relay Ventures in 2008 with John Albright. Kevin holds an MBA, with distinction, from York University, a BA degree in Strategic Studies, and the ICD.D designation from the Institute of Corporate Directors' Corporate Governance College at the Rotman School of Management, University of Toronto. Kevin also held an appointment as Adjunct Professor of Entrepreneurial Studies at the Schulich School of Business (York University) in Toronto. Kevin is based in Menlo Park, California.

On the planet today, there are more mobile devices than people. Approximately one-third of those devices are smartphones, and that number will grow as older technology transitions out of use. We are entering an era where, for the first time in history, every human on the planet will be addressable by machine.

When we think about mobile, we're not talking about the lone device in your hand. We're talking about all of the connected devices that we interact with on a daily basis. With a supercomputer in every person's pocket, the modern mobile age is exciting and empowering. Billions of people are going online through mobile devices, and not just in the developed world, but across the globe.

This new state of hyperconnectivity will profoundly impact society, technology, health care, education, commerce, and finance. With so many aspects of human life being affected, you might think this change would happen gradually, but that is not the case. The transformation is taking place at an unprecedented pace, making mobile the biggest and fastest-growing man-made platform in history.

Behind this revolution, there are four converging trends that act as the catalyst for change. The first trend is worldwide Internet usage. When it was created as a reliable communication system for the government back in the 1960s, the Internet was not meant to be a consumer product. Yet, it has been adopted into the mainstream in a way that has completely changed and expanded communications, and its influence can be seen everywhere. For example, traditional media has grown and adapted around the Internet, with a progressive blurring of lines between the Internet and television, radio and publishing. Much of what we do is now inextricable from online technology.

Back in 1995, people believed that the Internet would be transformational and, with luck, one percent of the population would be connected. We couldn't foresee the astronomical presence the Internet has in our lives today. At the time of publication of this book, there are more than 3.3 billion users on the Internet, representing almost 46 percent of the world's population. The number of internet users has increased tenfold from 1999 to 2013. The first billion was reached in 2005, the second billion in 2010, and the third billion in 2014.

The Internet has become so ingrained in our society that it's starting to be regarded as a human right alongside drinking water and clean air. I believe this is reasonable—the Internet is now the primary way we connect, share information, and access the collective intelligence. People who lack the means to get on the Internet are at a significant disadvantage socially, educationally, and economically.

The second trend is the fact that global smartphone adoption is what is going to drive the next three billion Internet users. More than ever before, we're able to access the Internet through apps and untethered mobile devices. People who are connecting online for the first time will take a different on-ramp than that of many longtime Internet users. The classic desktop computer, keyboard, mouse, and browser experience is already beginning to go the way of the fax machine. These three billion new users will skip the computer experience entirely and access the Internet through thirty-five-dollar Android smartphones. What's the impact when, in a few short years, 80 percent of the world's adults have a smartphone in their hands?

There are several results that we can confidently predict. The most obvious outcome is that widespread possession of mobile devices will drive greater Internet usage. We also know that there is a correlation that shows greater Internet usage the lower a person's income. Not only does this mean we'll have more people connected to the Internet, but we should also see more online engagement. We've never been less alone than in today's digital world, and interconnectivity is going to become that much more important as we move forward.

The third factor for change relates to the Internet of Things, or how all of our devices connect with one another. The smartphone era was all about the aggregation of sensors and creating multipurpose devices. When you think about the first mobile phone that gained widespread adoption, it was an analog device with a singular function: voice calls. One of the first features added to that phone was a camera. It was an awful camera that took relatively low-quality photos, but it was the first sensor integrated into a phone that wasn't related to making phone calls. And today, smartphone cameras have virtually replaced point-and-shoot cameras.

Over time, more sensors were added to the phone to serve

different purposes. When the first iPhone was introduced, it contained sensors such as an accelerometer to tell the user which way the phone was oriented and an LED light sensor that could brighten the screen when needed. Eventually, the advent of multiple imaging sensors allowed the phone to have both front- and back-facing cameras. Now we've entered an era where it is almost a competition among tech companies to see how many sensors can be put into a device. For example, the modern smartphone today contains nearly twenty sensors including gyroscope, accelerometer, magnetometer, proximity, fingerprint, temperature, light, pedometer, and of course at least two cameras. The smartphone is a marvel of technology that now vastly outranks the computer that put the first man on the Moon in the 1960s.

The Internet of Things is the opposite of the multisensor device—it's a fragmentation of the sensor bed. With the cost of sensors dropping to the point of relative insignificance, we can now have single-purpose sensors doing different jobs. A smart home could contain several of these devices, each performing a specific job but networked together and connected through the cloud. For example, a smart thermostat is basically a computer controlling the environment's temperature. A connected home could also have door locks that open when sensing a person's presence or the tools to change the color of connected light bulbs. The result is a system of single-purpose sensors that are combined into a network to perform specialized functions. The idea of low-cost sensors talking to the cloud is what defines the Internet of Things, and this is made possible by the supply chain of low-cost components designed for, and feeding, the smartphone ecosystem.

Predictions claim that, by 2020, we're going to have over six and a half connected devices per person on the planet, which adds up to approximately fifty billion more connected devices. People immediately think of smartphones when they hear the phrase "connected devices," but this category includes tablets, smart watches, home automation systems, smart televisions, connected cars, and more. Many of us in developed countries already own at least six of these devices. I believe the predictions may be wrong in that people will have *more* than the estimated six and a half connected devices per person. The numbers add up quickly, and I think most people will suddenly realize that they have more connected devices than they originally thought.

The fourth driver of change is the millennial generation. They are currently the largest generation, outnumbering the baby boomers and making up the majority of the workforce. In the next few years, they will be making 50 percent of purchasing decisions. Why is this important? The millennial generation was born and bred on mobile technology, and they are mobile natives. They use technology in a different way than people who adopted it later in life. This means that marketing and product design decisions will increasingly be made with millennials in mind, and companies will have to consider their unique needs and habits when developing new devices. The millennial generation is not unlike the three billion new Internet users that we'll see soon, in that many millennials skip traditional access to the Internet, instead coming online straight through mobile.

By 2020, we should see two to three times more smartphones

than PCs being used by consumers. The mobile smartphone is the first tech product to be bought by everyone on Earth, across multiple generations. A baby boomer may not know how to use all of the features of a smartphone, but the point is that this is the first device in history that was not made just for the technologically savvy. Computers require specific skills to operate, troubleshoot, and repair, but the smartphone is built to be intuitive and consumer-friendly. Everyone with a smartphone can figure out how to work the basics.

It's ironic that we have these infinitely more sophisticated devices in our pockets, yet we don't have to explain to everybody how they work. If you hand a two-year-old a tablet, he or she can poke at the screen and intuitively figure out how to navigate the device. The same can be said for an eighty-year-old grandmother who can take a photo and share it with her grandchildren. The process has become so easy and clearly explained that no one even realizes they're using a computer anymore.

These four trends—worldwide Internet usage, global smartphone adoption, the Internet of Things, and the millennial generation—will continue to shape the mobile revolution as it grows. Back in 2008, today's mobile reality was not a foregone conclusion, and we can easily say the same for tomorrow's. Mobile technology lives up to the title of biggest and fastest-growing man-made platform in history and shows no signs of slowing down. While we have an abundance of well-informed predictions, we can never know for sure how the future will unfold. What we can do is influence how mobile evolves.

As we move forward, it's important to understand how technology has evolved so we can remind ourselves not to hold too tightly to our definition of mobile today. Yesterday, mobile devices were used on the go. Today, they're also used in the home. Tomorrow, they will be used everywhere. We have worldwide adoption of devices that are more powerful, more user-friendly, and more ubiquitous than any other technology product in history, and they're all connected. Therein lies the massive opportunity for today's thought leaders, innovators, and creatives to shape the future of technology.

ARTIFICIAL INTELLIGENCE: THE BRAIN BEHIND THE MOBILE REVOLUTION

Bill Mark

BILL MARK leads SRI International's Information and Computing Sciences division comprised of 250 researchers in four laboratories creating new technology in virtual personal assistance, computer vision, machine learning, information security, and speech analytics. In addition to leading-edge research, the group has a strong focus on commercializing technology, licensing technology to corporations, and creating spin-off companies such as Siri, Tempo, Kasisto, HyLites, and Desti. Prior to joining SRI in 1998, Dr. Mark headed the System Technology Group at National Semiconductor, focused on system-level design and implementation of the silicon-based systems of the future. His previous positions include those at Lockheed Martin Palo Alto Research Laboratories; Savoir, a company developing software tools for flexible manufacturing, of which he was cofounder; and the University of Southern California Information Sciences Institute. Dr. Mark has a PhD in computer science from Massachusetts Institute of Technology. His personal research interests include virtual personal assistance and mediated spaces.

Artificial intelligence is at the core of many of the innovations and technologies that we use today. It is at work in our search engines, music recommendation software, and personal assistants, and it is learning and constantly improving. Whether it is visible or not, we interact with artificial intelligence every day. That is why when people ask how many years it will take before we see a profoundly intelligent system, they are mistaken: the technology is already here.

Research in artificial intelligence is, and always has been, the exploration of the phenomenon of intelligence. To understand the current state of advancement, it helps to look back at how the technology came to exist. The goal of creating artifacts that exhibit intelligent behavior has been actively pursued by people for centuries. Humans have been thinking about intelligent constructs for even longer, with examples of automatons appearing in myths that go back thousands of years. In the present sense, the technology started with the birth of modern computers in the 1940s and 1950s.

Before computers, autonomous intelligent creations were just a concept. It was only with modern computational technology that we were able to automate the learning and reasoning needed for artificial intelligence. Computers provided the necessary tools for scientists to turn the concept into reality. Alan Turing's work in the 1950s helped to ignite the scientific community's interest in computers that could mimic human intelligence. He developed the Imitation Game, which would later come to be known as the Turing Test, as a way to prove a system's intelligence.

For a system to win the game, it needed to convince a panel of judges that it was not a computer. It would have to, effectively, pass as human.

Though no system at the time was able to win the Imitation Game, researchers and innovators were inspired to create one that could. In 1956, a conference took place that included John McCarthy, Marvin Minsky, Alan Newell, Herbert Simon, and other scientists that marked the start of artificial intelligence as a research field. Since that time, the field has been growing constantly, if not at a steady pace. Advances are made in fits and starts, and like any other major scientific discipline, artificial intelligence as a field is split into subspecialties. In the early days, there would be a single conference covering subjects such as machine learning, perception, and natural language processing. Today, the natural growth of the field has resulted in multiple topics within artificial intelligence in which a researcher might opt to specialize.

Regardless of the subfield, when we talk about intelligence, we naturally think about it in human terms. The crux of the science is artificially to replicate the behaviors that we consider to be intelligent. One of the most important and uniquely human behaviors is natural language understanding—the remarkable ability we as humans have that allows us to communicate thoughts to one another in dozens of languages. The smartphone has been pivotal in bringing this technology to the consumer market, particularly in spoken form. When you use your voice to ask questions, dictate text, search the Internet, or use a program, you're using automated speech recognition and natural language understanding—two subfields of artificial intelligence.

Automated speech recognition is very high profile these days. The proliferation of smartphones means that speech recognition has vast potential to impact consumer markets directly. The technology is already at the point where Siri or another virtual personal assistant can understand what you say to it most of the time. However, the accuracy drops significantly when the user is in a noisy environment or moves away from the microphone. Comparatively, humans are decently capable of discerning speech in loud environments. Current research is directed at these types of scenarios in an effort to close the gap between human and computer performance.

Human natural language understanding is incredibly nuanced and complex, going far beyond current artificial intelligence capability. But for some tasks, artificial intelligence can match or even outperform a human. We have systems that can identify objects, activities, and environments in a manner akin to visual perception. This technology is used for purposes such as image recognition and self-driving vehicles. Other systems are designed to project their understanding of the world into the future to determine a course of action or react to unexpected events. Leveraging the machine's computational power, artificial intelligence programs perform tasks such as risk analysis and stock market predictions using vastly more data than a human can comprehend.

How can machines discover, remember, and use information about the world? This is the province of the subfields of machine learning and knowledge representation. Without much conscious effort, a human is able to process and categorize the objects and actions in their surroundings. Consider seeing a dog walk across a street. Simple, yet the brain has to identify both the action of walking and the dog. Correctly understanding that scene involves a whole web of interrelated pieces of knowledge: awareness that objects of a particular shape are dogs, identifying the street as part of the physical environment (and not part of the dog), and so on. These are things that we learn over time, from both personal experience and shared knowledge. Creating processes for gaining and connecting pieces of knowledge in a computer is what defines the field of machine learning. Storing the knowledge in a way that allows it to be recalled and used when relevant is what defines the field of knowledge representation.

The ability of a system to learn significantly expands the scope of its capabilities. Keeping with human terms, an apt analogy might be to think of a system using machine learning as a three-year-old child. It's true that we explicitly teach children certain things, but much of what they learn is gained from observation and personal experience. A child will, for example, see dogs over and over, until they are able accurately to identify new dogs. Repetitive exposure is the same process that researchers have used to teach artificial intelligence systems to identify photos of objects—this time using computer speed and ever-improving algorithms. The better the algorithms, and the larger the quantity of representative data, the more accurately the computer will be able to learn to identify the species. In the same way that a person's worldview develops from their experiences, artificial intelligence is shaped by the data we feed it.

Some of the most important knowledge that we learn through personal experience and shared knowledge is what we sometimes call "common sense." This is the kind of knowledge that allows a child to understand that, if he or she drops an object, it will fall. Common sense is the phrase we use to describe our innate understanding of the world that we naturally develop from experience. Some things we "just know," and because it is hard to explain exactly *how* we know them, and because we know so much, this type of knowledge is difficult to represent in an artificial intelligence system. Knowing that that is a dog walking across the street requires a great deal of common sense knowledge: a dog can move, streets can support the weight of animals, and on and on.

The different subfields—visual perception, natural language understanding, planning, machine learning, and knowledge representation—meet in the field of robotics. A robot is the physical embodiment of artificial intelligence. As it interacts with its environment, it needs to perceive its surroundings, plan how to move within that space, and respond to natural language and other input. A course of action is then determined by coordinating the feedback from different sensory inputs. The integration of all of these capabilities in one system makes it more challenging to design, but the complexity also allows for a deeper exploration of artificial intelligence's possibilities.

As the field moves forward, the goal of creating systems that exhibit intelligence remains at the core. However, that overarching goal can be broken down into three objectives that are necessary for progress. The first objective is to improve the technology's ability in core capabilities, such as classifying objects. We are already using artificial intelligence for these tasks, but to get better results, we need to refine and build upon the existing algorithms and data structures.

Another objective and major direction in the field is to develop systems that are better at understanding and interacting with people. The end goal of artificial intelligence is to make human life easier and more enjoyable, and to achieve that, we need autonomous systems that can understand us. Current systems, such as Siri, can already listen to us and often respond with a reasonable answer. Siri is able to understand questions and deduce the appropriate response, but there are limitations. It's difficult to have an extended conversation or get answers to highly nuanced questions from the system, but these are areas where we are making progress. This includes developing ways for the system to assess and appreciate the emotional state of a human during an interaction. Emotions are a critical part of human communication, and determining whether a person is feeling frustrated, confused, happy, or any of the other complex psychological states he or she may be experiencing will allow the system to provide a better response.

The third major objective is to combine research progress in all of the subfields in order to improve the practical applications of artificial intelligence. For example, a product recommendation system uses machine learning to select products, but its interaction with the end user is critical—its success hinges on how the user relates to the recommended products. This is because there is often no objectively better recommendation when the system has to decide between two products. To account for this, the artificial intelligence needs to interpret how the human reacts to its recommendation and adjust its future actions accordingly.

Researchers in the artificial intelligence field are actively making progress toward these three objectives, but unlike counting down the days until a product launch, there is no definite timeline leading to a point when the technology is "done." We are not progressing linearly, but rather making continuous advancements in a multidimensional space. If we look at Siri again, the system is quite good at some tasks and, in turns, awful or hilarious at others. It's not as though the system is half finished; some aspects are already in a sophisticated state, while others are in early stages but still in need of more refinement.

Even as we continue to advance artificial intelligence, it is already incredibly prevalent in our day-to-day lives. Any time you search the Internet, you're tapping into a system that has learned what to return for a given keyword, and that even knows something about what *you* specifically might be trying to find. I believe that as the Internet of Things grows and an increasing number of everyday objects possess computational and networking capability, artificial intelligence will grow with it. Intelligent systems are already such an integral part of computing infrastructure that it is unlikely one would expand without the other. For example, as the insides of our cars, the rooms in our houses, and our offices become more computational, there will be opportunities for artificial intelligence to improve our effectiveness in those spaces, helping us to interact with the devices and with each other. Along with this increasing presence of machine intelligence, we will also see a trend toward autonomy.

Autonomy for intelligent systems naturally comes with certain risks and challenges. If our autonomous system learns from users, we want it to pick up the best traits of humanity rather than the worst. For all autonomous systems, it's important to install safeguards that prevent bad behavior. However, it is not only system developers who need to consider possible roadblocks, but all of us. Whenever I discuss the challenges of artificial intelligence, I like to ask people to name their favorite human-computer interface. People give various answers, but then I say, "What about the automatic braking system in your car?" When you put your foot on the pedal, modern braking systems interact with a computer system to bring the car to a stop. I think most of us trust our brakes or, just as likely, never even think about them. Taking the technology a step further, would you trust a driverless car? The problem is that the public needs more experience with driverless cars and autonomous systems before trust can be established. These are systems we're going to depend on, and we need to develop mechanisms for understanding and trusting them, just as you would need to understand and trust another human driver.

Hand-in-hand with autonomy are the same morality

challenges that we face as humans. Looking at driverless vehicles again, we have some classic philosophical questions to answer. For example, if a car is on a collision path, the artificial intelligence will need to make a split-second decision. Does it brake, possibly killing five people, or does it continue its trajectory, definitely killing one? Does the car swerve to avoid a child, even if it means hitting three adults? These are the types of moral dilemmas that will have to be decided and programmed into the system. That said, they are the same considerations that any human driver faces every time he or she gets behind the wheel of a car. The difference is that an artificial intelligence system reacts to a moral dilemma quickly enough to have a choice, whereas a person reacts on instinct.

With so many areas in which artificial intelligence can be applied, you might wonder how it will affect us the most in the near future. I believe that there is enormous potential in the medical field for intelligent systems to enhance the way we diagnose illnesses, monitor patient conditions, and administer treatment. We can expect to see this manifest in the form of networked devices such as pacemakers, and ingestible nanobots that will move through our systems to find damage and repair tissue. In the near future, we'll also likely use artificial intelligence to assist aging populations. While the assistance may have a medical or functional component, it may also include social capabilities that could have an equally important impact. Whether the function is simply providing an entity to talk to, encouraging healthy activities, or reminding patients to take their medication, social interactions are a critical part of life that much of the aging demographic is sadly lacking. We don't only want to prolong people's lives with new technology, but to improve their quality of living as well.

In many ways, artificial intelligence has already improved our world and will continue to do so. The technology will become more humanlike, but artificial intelligence will be different than human intelligence—which is not necessarily a negative outcome. When you interact with people from a culture that is different from your own, their manner of thinking and habits may also differ, and I believe it will be the same with artificial intelligence. Computer systems will, in a way, have their own character and culture that will make our interactions with them a wholly unique experience.

Humanity has been dreaming of intelligent machines since time immemorial. As a staple of science fiction, the subject fascinates us, but the reality is that the technology is here today. Underlying the most exciting projects in the mobile era, from driverless cars to advanced robotics, is the foundation of artificial intelligence. As research takes us further down the continuum of progress, advancements in communication, problem solving, and machine learning will impact the entire field of technology. We're creating machines that, for the first time in history, have intelligence that potentially outranks our own, the implications of which are thrilling and, for some, terrifying. Regardless of how you feel about artificial intelligence, there is no denying that it opens the door to a world of possibilities. It is an active force in our lives today and will, undoubtedly, power much of tomorrow.

THE CREATIVE DESTRUCTION OF MEDICINE

Eric Topol

ERIC TOPOL, MD, is Professor of Genomics and holds the Scripps endowed chair in innovative medicine. He is the Director of the Scripps Translational Science Institute in La Jolla, California. Previously, he led the Cleveland Clinic to its number-one ranking in heart care, started a new medical school, and led key discoveries in heart disease. Eric is the author of two books: *The Creative Destruction of Medicine: How the Digital Revolution Will Create Better Health Care* and *The Patient Will See You Now: The Future of Medicine Is in Your Hands*. Eric was selected as one of the 12 "Rock Stars of Science" by GQ and the Geoffrey Beene Foundation in 2009. He was elected to the American Society for Clinical Investigation, the Association of American Physicians, and the Johns Hopkins Society of Scholars. He was named Doctor of the Decade by the Institute for Scientific Information for being one of the top 10 most cited medical researchers. In 2012, Modern Healthcare ranked Topol as the most influential physician executive in the United States. He lives with his family in La Jolla, California.

Whenever a new, highly anticipated device is released, people line up outside stores to buy the first one. Not just tech aficionados, but the average American clambers to get his or her hands on the newest, fastest, most powerful piece of technology. Never before have we seen such a rapid rate of advancement and adoption when it comes to the creation of new devices. But, historically, widespread adoption of technology has been a slower process. If we define widespread adoption as one-fourth of Americans using the technology, it took electricity over forty years from the time it was available for public purchase to the time it had gained prevalence. Television, personal computers, and the Web took twenty-six, sixteen, and seven years, respectively, to achieve widespread adoption. Keeping in line with this trend of diminishing time frames, smartphones earned widespread adoption in only two years. A convergence is occurring between the rate at which we advance technology and the rate at which we adopt it, and it doesn't seem to be slowing down.

While other industries are forging ahead into the bright, digital future, the medical field is dragging its feet. The answer to closing that gap may lie with the smartphone. Not only is the smartphone the fastest-adopted piece of technology in human history, but its impact has been huge. These devices have changed our daily lives, with so much of what we do now connected to our phones. For many people, their smartphone is their primary device for communication and information access. It is untethered in a way that previous technology was not, in that it is always on and always with us, giving us nearly uninterrupted access to an endless supply of information. This constant connection to the Internet has affected the way we work, the way we entertain ourselves, and even the way we focus our attention. Studies have shown that, in only ten years, the average attention span dropped from twelve minutes to five. Perhaps the most extraordinary aspect of the rise in mobile technology is that the change happened in a relatively short time frame—it took less than a decade from the introduction of the iPod to the appearance of the smartphone, e-reader, and tablet.

With modern technology rapidly advancing and becoming so ingrained in our daily lives, we have to ask ourselves why the medical field is struggling to keep up. I believe that a primary factor in this problem is that many, if not most, of the individuals in medicine are not digital natives. They did not grow up with this technology, but rather adopted it—or failed to—at an older age. In the United States, fifty-five is a significant number: approximately 55 percent of physicians are over the age of fifty-five. When we consider that a certain percentage of these professionals may not have kept up with changing technology, we can see why there may be issues of resistance.

While the field of medicine is lagging somewhat in the adoption of new technology, there are digital features that will, in time, transform the future of medicine. The biggest impact will come down to six characters, the nucleotides of the genome: 0, 1, A, C, T, and G. When you have the genome of each person fully defined, you have the medical essence of the human being. It will allow us to customize and tailor treatment to the unique needs of each patient in a way that we've never done before. Complete access and an understanding of genomic data will unlock personalized medicine on a wide scale, but we have not yet reached that point.

Today, we practice medicine at a population level. Everything we do in medicine essentially fits into one of three purposes: screening, diagnosis, and treatment. When done at a population level, many procedures are imprecise and remarkably wasteful. One example of a screening procedure practiced on a large scale is mammography. One might think that if we screened every woman above a certain age, breast cancer would be identified earlier, resulting in more timely treatment and better outcomes. However, research—including a massive study looking at ninety thousand women over a period of twenty-five years—has shown that there is no difference whatsoever in outcomes with routine mammograms. Some studies have shown that this practice actually results in net harm—more women are affected by negative side effects of mammograms than those who are helped. Wasted resources and unnecessary procedures are an unfortunate and unavoidable outcome of impersonalized medical treatment, but the way we practice medicine is beginning to change.

Technology is ushering in a new age of personalization with the tools necessary to digitize human beings. We've already seen similar applications of technology used to organize and display complex data—for example, geographic information systems. You start with a map and layer information, such as traffic, street views, or landmarks, and integrate the layers to display multidimensional data. Now we can do the same with a human being. One layer of the human information system that we already have a wealth of data on is the social layer. Social media sites such as Facebook show an extraordinary amount of demographic information that can be gathered on any active user. The benefit of social technology is that it can bring people together like never before and is increasingly weighed toward mobile usage, which makes it easier to access. The downside is the compromised security and privacy that comes with storing so much personal information on one device, and the same risks will apply when it comes to mobile health data.

There are already many exciting applications to explore in the burgeoning field of digital medicine, some with the potential to transform the future of health care. We're seeing data capture and sharing move beyond purely social applications to more health-related areas, such as fitness. Wearables that use sensors to track physical activity are increasingly popular, with wireless accelerometers coming in as the front-runners of the digital wearable revolution. These devices commonly function as pedometers, sleep trackers, and heart rate monitors, but they are expanding beyond that to gauge posture, monitor teeth brushing, track how much the user is eating, and more. As the list of wearable functions grows, so does the sophistication of the technology used in the devices. Interesting applications include a digital mood ring that measures electrodermal response and heart rate variability to infer the wearer's stress level, and even a device that, when pointed at a food item, lists the ingredients through spectroscopy. The influx of new health-related lifestyle technology is impressive, but the accuracy of these devices needs refining.

On the true medical side of health-related devices, sensor technology is radically changing how we approach chronic conditions, both in terms of management and prevention. In my cardiology practice, most of my patients have hypertension and now wear wireless blood pressure devices that capture their vitals. There are several benefits to using this technology over the traditional tracking methods. The device is able to capture a wider range of data, giving a more complete report of the patient's health, and to send it to his or her doctor in a single screenshot. In the past, the de facto way to track a patient's heart rate at home was by using a Holter monitor, which was invented in 1949. Now, that's been overridden by a device the size of a bandage that captures every heartbeat for two weeks, without the wires and inconvenience of the Holter monitor.

Similar applications exist for monitoring glucose—patients with diabetes can monitor their glucose levels right through their smartphone. There are also several mobile devices that produce electrocardiograms and will give an immediate algorithmic reading. I used this technology with a patient and later received an email from the device notifying me that the patient was in atrial fibrillation. This level of connectivity is unprecedented in medicine and is already allowing doctors to provide a higher level of care for their patients.

By facilitating a more personalized approach to patient care, new technology is helping save lives. A study done in Australia monitored the use of a device in a drugstore that gave electrocardiogram readings simply by having a person place their finger on a sensor. The researchers found that of one thousand people who had no history of arrhythmia, fifteen were unknowingly in atrial fibrillation, which is a precursor for stroke. By using this device, those people were able to follow up with their doctors and possibly avoid a serious medical crisis. Imagine how many more strokes might be prevented if we were able to monitor individuals beyond spot readings.

There is a great deal of hype around watches, with entries from prominent companies such as Samsung, Qualcomm, and Apple promising the ability to report medical information. The evolution of the watch is interesting because people no longer buy digital watches with the primary purpose of telling time, but you can obtain one that will predict a seizure in someone with epilepsy before it happens. As opposed to the fairly inaccurate consumer lifestyle watches, there is a watch that, albeit being clunky, can track vital signs continuously and very accurately. This particular device has been FDA approved for use in hospitals and is able to provide ICU-level monitoring in every patient room, replacing the single visit by a nurse once per shift with continuous monitoring. Doctors are then able conveniently to access a patient's dashboard right from their phone or computer.

Going beyond wrist devices, there is a growing pool of wearables: necklaces that track cardiac output in people who have heart failure; socks that monitor vital signs of newborns or report pressure in diabetics with neuropathy or foot ulcers; contact lenses that measure eye pressure in patients with glaucoma or test the glucose levels in tears. There are also mobile devices that are not wearables, but that function similarly, such as floor tiles that help seniors who have balance issues to avoid falling.

Then there is breath, which gives an incredibly rich amount of medical information that hasn't been fully harnessed until the smartphone era. The most surprising and potentially impactful use of breath is to diagnose cancer. Certain organic chemicals, such as alkenes, are present in the breath of an individual with cancer. Given their high endowment of olfactory receptors, dogs are able to identify these chemicals through scent. However, training dogs to perform this task is resource-intensive and does not produce a consistently accurate solution. We now have devices that function as electronic noses, which can attach to smartphones to simulate a dog's ability to smell cancer. Breath and voice are both useful in gauging a person's stress level, as they both give indications of an individual's state of mind or mood. We're even able to use voice to diagnose Parkinson's disease with 99 percent accuracy.

The degree to which we're integrating medical devices with the human body goes far deeper than wearables. Technology such as edible sensors composed of a digestible pill containing a chip, sensors that are embeddable in the blood stream, and biodegradable chips that can be placed directly on the skin are all being developed. A microfluidic skin chip from the work of John Rogers at the University of Illinois Urbana is able to capture electrocardiogram, EEG, and EMG by way of muscle signals. While we can gather a great deal of information about the human body externally, it is particularly exciting to consider what we can capture internally with microtechnology.

Embedded devices can be lifesavers for people who are predisposed to illness or whose conditions require extra monitoring. For example, the deaths of Tim Russert and James Gandolfini via heart attacks may have been prevented. In their cases, each man had tested negative for artery blockages, when in actuality they had mild narrowing of the blood vessels. Stress tests only pick up tight narrowings that impede blood flow, so their problems went undetected. For several years, scientists and doctors have been collecting blood in the earliest minutes of a heart attack to find cells that are shed by the lining of arteries during an attack. When these cells are sequenced, their particular genomic signals can be coupled with an embedded sensor, which would alert the patient before a heart attack fully occurs. A sensor developed for this purpose at Caltech is smaller than a grain of sand and can be injected directly into the bloodstream. Had Tim Russert and James Gandolfini been embedded with this type of sensor, their artery blockages may have been noticed in time to take lifesaving action. This type of sensing isn't limited to heart attacks, but could be used to detect diabetes, tumor DNA in the blood, and more. The new applications of sensor technology are quite innovative and ingenious, but developing the prototypes is just the first step.

All new medical devices need to be tested and validated through rigorous clinical trials. Validating effectiveness is key both in terms of safety and in order to avoid situations in which a seemingly promising solution does not follow through on its hype, as was the case with the proton beam. However, if new digital medical technology can be validated, the potential exists to reduce our health-care expenditures by at least a third. These savings would come from areas that have long been icons of the medical field, such as hospitals, office visits, current lab procedures, and the way we perform physical exams, all the way down to the classic stethoscope. What we have conceived as the medical metrics of today may look quite different tomorrow.

Changes to the medical field may occur in a drastic fashion, the most noticeable of which will likely be the edifice complex. In 1946, George Orwell described hospitals as the anti-chamber to the tomb, noting that the basic hospital structure hadn't changed much over the course of medical history. With mobile technology, certain procedures that were once confined to the hospital will be available beyond its walls. The large buildings of our current hospitals may be whittled down and replaced with only the core areas of the intensive care unit and operating room. Less urgent or ongoing medical care could plausibly be done at the patient's home or on an ambulatory basis.

Moving patients out of the hospital has financial benefits and may also improve patients' well-being, as studies have shown that one out of four people is actually harmed in the hospital. The problem exists at an institutional level: nearly 2,600 American hospitals were rated by Consumer Reports on a scale of one to one hundred and, shockingly, the highest rated hospital was given a score of only seventy-six. The added risks and resulting injuries to patients are obviously exceptionally costly. There is also a huge time cost associated with our current medical system. The average wait time to get an appointment with a primary care doctor in the

United States is at minimum two weeks, ranging up to sixty days in densely populated cities. Not having timely access to medical care is a serious problem, and the industry isn't listening to patients. In a Cisco survey of thousands of Americans, 70 percent said they would prefer to have interactions with their doctor virtually, not physically. In areas where a virtual model has been tested, the results were positive. In a study by Virtuwell, a sizable Minnesota health insurer, virtual visits saved patients nearly ninety dollars on average due to the lack of unnecessary testing usually performed during in-office visits.

The hospital of the future is not a sleeker, flashier building, but the average home equipped with digital sensors and tools. Doctors are consulted virtually on demand. We've witnessed an Uber-ification of our economy, with transportation and other services available almost instantly through apps. In the Mission District of San Francisco, you can even have your medication delivered by a drone—for an added fee, of course. Everything is trending toward convenience and on-demand availability, and now we're seeing this change extending to doctor interactions. I believe that, in the not-so-distant future, visiting the doctor's office may be as obsolete as going to a video store. It will be an activity saved only for exceptional circumstances. We will always need physical locations to receive medical treatment, but not to the point where visits are nearly 100 percent in person as they are today.

Laboratory testing is another area desperately in need of improvement. I think everyone has had this experience: you go to the doctor's office to get your blood tested, the doctor sends the sample to the laboratory, and then you never hear back about the results. You either beg and grovel to get your data, or you never get it at all. Fortunately, we're beginning to see changes. Rapid drugstore lab testing, which gives results in minutes, is just the beginning because, with microfluidics, we can run tests through a smartphone. Thyroid, liver, and kidney function tests, for example, will all be available, as well as testing for infectious diseases such as HIV, tuberculosis, and malaria. With the technology validated and perfected, we should be able to run nearly any laboratory test through the microfluidics-smartphone mechanism. We can theoretically even gather genomic data through the smartphone, allowing us to sequence pathogens or test DNA for a particular allele denoting risk or response to a drug.

Imaging is yet another capability that may be possible with smartphones in the near future, allowing one device to replace a collection of traditional medical equipment. If someone falls and a fracture is suspected, they could take a picture with their smartphone, transfer it over the cloud, and receive an algorithmic determination of the X-ray results. You can use your phone as a powerful otoscope to examine a person's eardrum or perform eye exams that give the same results an ophthalmologist would get in their office, all on a smartphone for under one hundred dollars.

Smartphones may even replace the iconic stethoscope, a medical kit staple for nearly two hundred years. It has served well in the past, but now you can get a high-resolution ultrasound with a smartphone, which provides richer data and fits in your pocket much better than a stethoscope. Perhaps most useful of all is that the smartphone records the captured data and makes it accessible anytime, anywhere. As a cardiologist, I can now do a full echocardiogram in seconds and see everything that is going on with my patient's heart. The stethoscope is a relic that I haven't used in my own practice in years. The next generation of doctors and medical schools are beginning to wise up to the fact that digital is the future, but we still have a long way to go.

Historically, doctors have been conservative about putting new ideas into practice, so it is no surprise that there are problems with doctors getting on board with the major changes happening in their profession now. We have doctors who are still unwilling to give a copy of their office-visit notes to their patients, and we need to get over this roadblock to an area of greater transparency. On the other side of things, patients also need to get on board and become more engaged in their own health care. They need to advocate for themselves and press doctors to make digital services available. This wave of patient activation is occurring ever more rapidly as applications are developed for patients to generate their own data, connect to online health social networks, review their doctors, and more. A growing number of states allow patients legally to order their own lab tests without even visiting a doctor, and companies are cropping up to foster this. Both individuals and employers are fed up with the current cost structure and other problems embedded in the health-care system, and they are crying out for change. It is consumer empowerment in action, and I believe that consumer demand is the key to getting the medical community to adopt new technology.

Another critical component in the success of digital medicine is gaining the approval of regulatory bodies. The FDA has thus far been supportive of new technology and quickly approved the use of devices such as the electrocardiogram app and the embedded digital chip pill. We've also seen products start to hit the market and launch aggressive ad campaigns only to be shut down by the FDA, but this largely happens because the development team didn't work within the FDA's guidelines. It's necessary for companies to cooperate with the FDA, play by the rules, and adhere to the FDA's guidance if they are to be successful.

There is unquestionably a medical revolution brewing, with the smartphone as its tagline. Mobile technology is gaining power and will change the way we interact with doctors, diagnose illnesses, monitor conditions, and exchange data. We'll be able to get answers to our health questions faster than ever before. If you have a suspicious skin lesion, why would you go to a dermatologist's office when you can get it checked out virtually? Why would you visit an optometrist when you can use a smartphone attachment to get your glasses prescription? Why would you spend money on the emergency room to diagnose an ear infection when you can receive an immediate answer via the cloud? You probably wouldn't.

The mantra of the medical revolution is this: nothing about me, without me. There is no reason that test results and records should be shrouded in mystery any longer. It's time to involve the patients. In the smartphone era, people have the opportunity to be more involved in their own health care and become empowered with data, but we need to get medical professionals on board. It will be important that the digital doctor of tomorrow embraces change, adopts new technology, and respects patient-generated data. While it may be the most challenging time in health-care economics, it is also the most exciting time for potential advances in medicine. Let the digital revolution of medicine begin.

EDUCATION'S DIGITAL FUTURE

John Couch

JOHN COUCH is Vice President of Education at Apple. He has over forty years of experience as a computer scientist, executive, and advocate for technology in education. Prior to Apple, John was the Chairman and Chief Executive Officer of DoubleTwist, Inc., a provider of genomic information and bioinformatic analysis technologies. John originally joined Apple in 1978 as Director of New Products reporting to Steve Jobs. He was Apple's first Vice President of Software and Vice President/General Manager for the Lisa division. In 1985, he became Chairman of the Santa Fe Christian Schools turning the school into one of the first examples of how the creative use of technology can revolutionize learning in the classroom. John holds a bachelor's degree in computer science and a master's degree in electrical engineering and computer science, both from the University of California at Berkeley where he was honored in 2000 as a Distinguished Alumnus. He has also been awarded an Honorary Doctorate of Humane Letters and a "Leadership in Innovation Award" from Philadelphia University. He is the author of the textbook, *Compiler Construction: Theory and Practice* and has taught at both University of California at Berkeley and Cal State San Jose. He is currently working on an education book on new dimensions in learning.

Today's classroom should not look like the classroom of several decades ago, not just because of the tools that are available, but because of the students themselves. Students today are digital natives. They've grown up having unprecedented access to technological social environments and nearly unlimited information on the internet. This technological literacy has changed the way students learn and has altered the challenges that we face in the education system. While the classroom of today looks slightly different than yesterday's, tomorrow's classroom will exist on a whole new level of digital learning.

When we look back at education's first foray into the digital world, my thoughts start with Apple's introduction of the Apple II into the classroom. The Apple II was then replaced with Macs, and today we hear all about iPads in the classrooms. The education system is changing as technology advances, and I think both industries are ripe for a creative disruption. The downside of the education system today is that the pedagogy is still based on a consumption model. Students are taught about a process by moving it from A to B in a linear fashion, and then are told to memorize the information because the test is on Friday. The problem is that this method is not based on creativity. The student is consuming information, not creating, which is a huge missed opportunity.

Fortunately, the lack of creative opportunities in the classroom is a challenge in which technology can come to the rescue in a number of ways. Digital tools can help teachers significantly when it comes to managing data and time. At points in my career, I've been asked to do impossible things, and I would always respond, "I believe in miracles. I just don't believe in scheduling them." We're asking today's teachers to schedule miracles. Any given teacher will have a classroom of students who have a wide range of strengths and weaknesses. It takes a huge amount of time to diagnose each student, grade their assessments, and determine which learning activities need to be presented to each of those individuals to overcome their knowledge gaps. After that, the students are retested and regraded. When the teacher has finally completed this series of events, the time frame in which they need to be working on something else has come and gone. There is not enough time in the week for a teacher to meet the needs of each student. I believe this challenge can only be solved by determining the appropriate learning activities for each student through analytics.

The current student generation is poised for digital integration in the classroom. They're capable of working with a system that can capture the needed information, and their familiarity with technology is evident when we consider how the world looked as they grew up. Incoming college freshmen were children when Google was created. They've never had to walk into the library and go through the Dewey Decimal System to look up information. They do an automatic search, and everything they need is at their fingertips. These students were in elementary school when the ecosystem of music changed and you could easily download a song to your iPod or computer. They were just a few years older when Facebook and Twitter came into existence and widespread social sharing became popular. They were in middle school when the iPhone and App Store came out—many of these students have probably never walked into a retail store to buy an application because they're all delivered digitally. Ever since grade school, these students have been raised in a collaborative publishing mode.

It stands to reason that students who have had all their information and applications delivered to them digitally would prefer the same for their educational content. To meet students' needs, how do schools have to change? What will be the impact on education? We have this traditional paradigm in schools where the teacher is at the front of the classroom, telling the students what they need to know. What should learning look like?

We can start by examining how information is currently conveyed and gauge its effectiveness. A study done at the Massachusetts Institute of Technology measured the activity of brain waves during different activities. They observed that the participants' brain waves were passive while watching television. This result isn't surprising, since watching television is a fairly unengaging activity. However, when the scientists measured the student's brain waves while in class, they also found a passive level of activity. A teacher might look at the study and say, "Well, the student is sleeping in class," which is not the case. To the contrary, if you look at brain waves of someone who is asleep, they are quite active. The data shows that we need to rethink the lecture model of teaching, but change is not without challenge.

This generation of students grew up in a digital world and

look at technology not as a tool but as an environment. They want to create rather than just consume information, yet today's education pedagogy is based on the assumption that everyone is the same, everyone must jump through the same hoops, and everyone can learn by consuming. Instead, we ought to be strengthening the community of learning. Most schools today are high on content, but low on collaboration and virtually devoid of solving relevant problems. We don't ask students to solve problems but ask them to memorize. How do we instead put content in a relevant, challenging context?

To encourage critical thinking, we need to build an educational ecosystem that is relevant, creative, collaborative, and changing. Although the revolution has started, it's just not equally distributed. Back when I took chemistry in grade school, the periodic table was taped to the wall, and we were told to memorize it. Flash forward to the present, when I visited a one-to-one iPad classroom. The students were sitting at tables of three with a colored Easter egg in front of them. In the egg were colored beads, each representing a different element. Each group received elements that form water and were asked to determine whether their water was safe to drink. They were asked follow-up questions, such as, "In the unsafe water, are the elements carcinogenic? What percentage of unsafe elements are acceptable in drinking water?" This classroom activity wasn't an exercise in memorization, but a contextual problem requiring critical thinking. It's a different pedagogy than the mainstream education.

The biggest challenge in changing the teaching paradigm

is first to change people's attitudes. We need to start with teachers and help them come to an understanding that the nature of the modern student is to be actively creative rather than passively accepting information, and that, currently, we're letting a lot of students down by teaching to the least common denominator. Technology is able to help educators by capturing the data, providing analytics, and allowing them to prescribe the appropriate learning activities that an individual student needs. Once teachers realize that technology can be such a powerful tool, we should see many classrooms embracing a true digital learning environment.

Looking forward at the ideal school of the future, I believe that students will be more proactive in their own learning. We'll be able to determine a student's dominant learning style, whether it's visual, kinetic, or a combination of different modes, and be able to deliver material in the way that suits him or her best. Having access to data and analytics will allow us to understand where there are gaps relative to the given framework. With those understandings, we'll be able to deliver digital resources to a student to overcome their unique learning gaps. I believe we have the ability to create a learning environment that works just as well outside the classroom walls as it does inside the classroom today.

The impediment is the traditional institution and the old way of doing things. Currently, that's a one-size-fits-all model, but differentiating learning styles is not a new concept. There is a folklore saying, "Everybody is a genius. But if you judge a fish by its ability to climb a tree, it will live its whole life believing that it is stupid." We're asking all of our students to climb the same tree because we don't have the baseline of where a particular student is at a given point.

The times when we recall information the best tend to be the times when we're more creative, and devices bring creativity into the classroom. Ultimately, I believe pressure for change will come from parents. Parents will be looking for learning environments that meet their children's individual needs, and when they find the right tools, I think we'll see them advocating for those environments to be present in the classroom. Schools might not recognize the direction education is going, but there are companies and technologies being developed now that challenge the traditional institution.

Massive open online courses (MOOC) are an example of new educational environments. They allow for opportunities that simply don't exist in the traditional classroom. Hypothetically, there's nothing stopping a nine-year-old child from taking Stanford courses online, and the accessibility of these courses has undoubtedly contributed to their enormous popularity. There are online classes on Khan Academy, Coursera, iTunes U, and a number of other places that have hundreds of thousands of people enrolled. The wide availability of information that these online classes offer is incredible, but I think we have to be careful. If the pedagogy doesn't change—if it's still based on consumption and doesn't allow individuals to solve problems creatively—then these online classes are not going to work.

Technology isn't only affecting the way students learn, but it's also changing the business side of education. Current students—digital natives—aren't going to be hired to look up information. They are going to be hired to solve problems. They have access to experts all over the world for that. For example, when my son was in tenth grade, he came to me and said, "Dad, I need to do a science project." I asked him what he was interested in, and he said, "I read in a newspaper about deformed frogs—frogs being born with multiple legs. I'm interested in that." I knew his teacher wouldn't know anything about the topic and that he wouldn't find many books or scientific articles in the library either. This was in 2001 when the Internet was new, but when I asked him where he planned to find resources, he said, "I'm going to find information online."

To my surprise, when my son came back to tell me what he'd learned, he had a lot to say. "The frogs' issues are caused by three possible factors: ozone depletion and ultraviolet light, pesticides, and a parasite that imbeds itself to the hind legs of frogs." He had found a short article about the parasite and decided to contact the author, a professor in New York, for more information. I was still skeptical that he'd make much progress, but sure enough, he got in contact with the professor and met up with him in Oregon over the summer. Being a digital native, he had his MacBook and digital camera and was able to document the whole process as they studied these parasites. When he returned home, he wrote up his paper on the subject and ended up winning the science fair. Allow me to remind you that he was in tenth grade. When asked by a Stanford professor if he wanted to continue work on his project during the summer, he said, "No, I play basketball during the summer."

My point is that we don't give kids enough credit for how capable they are. However, they seem to handle contextual problems better when they're able to think beyond the linear processes that they're taught. Apple developed a framework called challenge-based learning, which essentially allows kids to pick a big problem and solve it. Ironically, the students who had the most difficulty with challenge-based learning were the honor students. They had grown so used to looking at a problem, categorizing it, and applying a formula that they weren't accustomed to thinking outside the box.

One of the first challenges that Steve Jobs gave me at Apple involved the educational ecosystem. He said, "All books, learning materials, and assessments should be digital and interactive, tailored to each student, and should provide feedback in real time." That's the ecosystem that I think we need to build, in the same way that we built an ecosystem around music that allowed us to purchase, collect, and play music digitally. We've built an ecosystem around the iPhone with the App Store that allows us to modify our phones to be whatever we want. When I look at Apple over the years and the different industries we've transformed, I think learning is the obvious next frontier.

Technology can be used as a substitution, or it can be used as a transformation vehicle. We've seen this over the decades as we shifted away from making a trip to the computer lab to having access to the world twenty-four hours a day. Transformation means making something possible that you could not do without technology, such as analytics on a large, in-depth scale. It's collaborating across space with someone on the other side of the world and sharing information. It's going beyond memorizing and regurgitating the content. Knowledge is not a commodity that is delivered from teacher to student, but something that emerges from a student's own curiosity-fueled exploration. How do we encourage that creativity? By embracing education's digital future. As E. C. Wilson said, it's not about IQ; it's about curiosity. It's time we build a learning environment that empowers curiosity and rewards innovation.

TRANSFORMING MOBILITY WITH SELF-DRIVING CARS

Jennifer Haroon

JENNIFER HAROON is the Head of Business, Self-Driving Cars at Google[x], where she focuses on strategic initiatives, business operations and partnerships. She previously launched Project Link, a wholesale metro fiber network in Kampala, Uganda, represented Google on the Alliance for Affordable Internet (A4AI) and launched a number of Google's health-related initiatives, including Google Flu Trends and the Flu Vaccine Finder. Prior to Google, Jennifer was a Project Leader at the Boston Consulting Group. From the San Francisco and Stockholm offices, she worked with organizations ranging from global public health to private equity. Jennifer also worked as an Associate in the Equity Research group of Thomas Weisel Partners. Jennifer received her B.S. from Duke University and MBA from UC Berkeley's Haas School of Business.

In 1885 when Karl Benz invented the automobile, the inventor unveiled his vehicle to the public and promptly crashed it into a wall. Ever since that first crash, automakers have been on a quest to make cars safer and smarter. We've made them stronger and better able to survive impacts. We've added safety features such as seat belts and airbags. Driver-assistance technologies such as automatic braking or lane guidance can reduce crashes, but all this is a workaround to compensate for the largest cause of fatalities on our road today: human error.

Cars are a critical part of daily life for millions of people all over the world. We rely on them to get to our jobs, to visit family and friends, and to live our daily lives. But no matter the make or model, they all share one common flaw: the driver. With self-driving cars, we can make driving safer and change the way we get around. Over one million people die on the world's roads every year, with over thirty thousand fatalities in the United States alone. Ninety-four percent of those fatalities are caused by human error. To put it in perspective, that's roughly the equivalent of a Boeing 737 airliner falling out of the sky every working day. With so many needless accidents and the general stress of traffic, it's no wonder that many people would prefer to do just about anything but drive. Unfortunately, this is where distracted driving comes into play.

With mobile phones, it's more tempting than ever for people to operate their vehicles without devoting their full attention to the road. Many people don't realize that distracted driving is such a harrowing problem, but it is the primary cause behind 60 percent of moderate-to-severe crashes, as reported by the AAA Foundation for Traffic Safety. People may not realize the effect of reading a short text message while driving, but when you take your eyes off the road for just five seconds when going fifty-five miles per hour, it's like driving the length of a football field blindfolded. Regardless, many people don't equate the risks of distracted driving with those of a clearly ill-advised action such as driving blindfolded.

While the risks may be underestimated by the average driver, most people still know that distracted driving is dangerous. It begs the question, why are people so distracted behind the wheel? One factor is that distractions, via mobile phones, are more accessible than ever before. Another is that traffic is actually getting worse. Over the course of two decades, the total vehicle miles traveled in the United States has increased by over a third. Meanwhile, the number of road lanes has increased by only a fraction of that. Our infrastructure is simply not keeping up with the rise in traffic, and the resulting time spent sitting on the road has a very human cost.

The average American worker spends fifty minutes per day commuting. When you multiply that figure by 120 million workers, it adds up to nearly six billion minutes wasted every day. Based on the average life expectancy, that's the equivalent of 162 lifetimes spent every day trying to get from point A to point B. Self-driving vehicles have the potential drastically to cut down on time spent driving and on traffic, not to mention the life-changing effects they can have for people who aren't currently able to drive at all.

One individual who tested Google's self-driving cars is an incredibly capable man, who is blind, named Steve Mahan. Based on where he lives and works, it should take Steve roughly half an hour to drive to work. Instead, it's a two-hour ordeal of patching together public transit or asking friends and family for help. When Steve was given the use of an autonomous vehicle for a few hours, he didn't go anywhere out of the ordinary. In fact, the things he chose to do might be considered by most people a bit mundane—picking up dry cleaning, getting fast food at the drive-through, running errands—but for Steve it was anything but routine. These everyday activities would normally be much more challenging or time-consuming for him to accomplish, but with the help of a self-driving car, he was given the type of freedom and independence that he often lacks.

Senior citizens are another demographic that could benefit significantly from the increased independence afforded by self-driving vehicles. In the United States, the majority of seniors over the age of sixty-five live in suburbs or rural communities where they are dependent on cars. That reliance puts them at significant risk for social isolation when declining health requires them to give up their car keys. The ramifications of isolation are not to be underestimated—studies have shown that seniors who are lonely are 50 percent more likely to die early than those who have regular social interactions. Independence is such an important factor in quality of life, and there are many others for whom a self-driving car could transform their mobility.

The benefits of self-driving cars are clear, but how can we address their challenges? Many automobile companies and individuals believe that, over time, incrementally adding driver-assistance technologies—features such as automated parking and collision-avoidance braking—will eventually lead to a fully self-driving car. From what we've seen, that is not the case.

Adding assistive technology centered around the driver will not lead to a point where the driver is unneeded, because the transition from driven to self-driving is not a continuum. It's a chasm. To put the concept in simple terms, it's like saying that if I work hard at jumping, one day I'm going to fly. It's not going to happen. There are three main issues that contribute to this chasm, the first of which I've already touched on: driver awareness. Driver awareness is already a problem, and with driver-assistance technology it is not being solved. In fact, we've seen that it actually makes the issue worse. Therefore, there is more needed for a self-driving car to be successful.

An alternative paradigm is the approach being taken by Google. We ran the first autonomous vehicle trial with human passengers using approximately one hundred Google employees. The participants were trained and, while we informed them that the car worked well, we made sure they understood that it was a prototype, and they needed to pay attention. As the trial progressed, we received the type of feedback that any product team would be thrilled to hear. Our participants said the car was amazing, that it was life changing, and that not dealing with traffic on a daily basis had a positive impact on their lives. However, when we took a look at what people were doing in the vehicle, we saw a few behaviors that caused concern.

In one anecdote, there was a gentleman driving on the highway at sixty-five miles per hour, when he noticed that his cellphone battery was running low. He turned to the back of the car and rummaged around in his belongings, finally pulling out his laptop and setting it up on the seat next to him. Then he turned around *again*, this time retrieving a charging cable and plugging his phone into his laptop. All the while, he was supposed to be paying attention to the road in case he had to step in. Based on our observations, it would seem that as technology gets better, the human driver becomes less reliable. Therefore, we weren't convinced that making cars incrementally smarter was going to get us the positive impacts we were seeking.

In addition to driver awareness, the second issue to consider is the car's ability to avoid collisions. Currently, we have cars that fully rely on the driver for control. They aren't going to malfunction and hit the brakes when they aren't supposed to, but they also aren't going to help you avoid an accident. We also have cars with driver-assistance systems, which still rely on a human driver, but they are also able to step in and help you avoid some percentage of collisions. In the United States, the average driver makes a mistake that leads to an accident once every 100,000 miles. Therefore, the driver-assistance technology needs to take action once every 100,000 miles to avoid a crash.

When we look at a fully self-driving car, the number of actions the system needs to take is approximately ten per second, or one thousand per mile. That's different from driver-assistance technology by eight orders of magnitude—the same as comparing the speed at which the average person runs to the speed of light. For a system to function at this degree of precision, it requires significantly more advanced hardware, sensors, and software than driver-assistance technology.

The third issue in creating a self-driving car is how the system deals with uncertainty. The system has to be able to make judgments about everything it senses. Is that gentleman about to step into the road, or has he reached the edge of the sidewalk and is now going to stop? It's incredibly difficult to tell. In a driver-assistance system, the car doesn't need to make that call because it has a human driver. In the case of a fully self-driving system, the car has to recognize that the situation is uncertain and take proactive action, such as gently braking, to avoid an accident.

At Google, we've explored these challenges and designed a working system. There are several key functions needed for the car to operate successfully. First, the car needs to know where it is in the world, which it determines by using a map and data from its sensors. It also needs to process what it sees in the moment and be able to differentiate between objects. For example, our self-driving cars can identify whether an object is another vehicle, a bicycle, a pedestrian, or something situational, like a traffic cone. Finally, it needs to predict what is going to happen and react accordingly.

It's not enough to just know what one other car is going to do—the system needs to analyze the whole situation and predict everything that might happen. Making predictions involves some incredibly complex processing, but once the car has captured the entire scene, it needs to respond in the moment by determining trajectory and speed. When we first tested self-driving cars, we restricted them to the highways. This setting was simpler, as all cars headed in the same general direction and the car could process a geometric understanding of the world. Once we started driving on streets, it introduced a whole new level of complexity. The car had to manage pedestrians crossing in its path, cars going in all directions around it, crosswalks, streetlights, and more.

There are so many nuances the system needs to be able to pick up on, as well as hypothetical situations it must be prepared to encounter. For example, in construction zones, the car has to recognize that other vehicles will act unusually. They might drive outside of the street lines or respond to someone using arm gestures to direct traffic. There may be police cars, and the system needs to recognize that a vehicle with flashing lights on it is not any old car. The same applies to school buses and other vehicles with special rules. After the system identifies these cues, it needs to know that other cars have expectations of it and then react appropriately. So when a cyclist puts up his arm, it means he expects the car to yield to him and make room.

Part of how we accomplish all of this is through sharing data between vehicles. For example, when one vehicle comes across a construction zone, the other cars learn about it so they can be in the appropriate lane to avoid some of the difficulty. Sharing data between cars is useful, but our understanding is actually much deeper. We can use all of the data that our vehicles have seen over time, the hundreds of thousands of vehicles, pedestrians, and cyclists they've encountered, to understand how things look, and then infer from that how they *should* look. More importantly, we can take a model of how we expect vehicles and people to move through the world and learn from that.

Predicting how traffic will react is critical, but there will always be things we encounter that we've never seen before, or that we see so infrequently that we haven't learned behavior for that situation. For example, one day our vehicle was driving around town when it encountered a duck in the middle of the road, and that duck was being chased by a woman in an electric wheelchair wielding a broom. Nowhere in the DMV handbook does it tell you how to deal with that situation. Luckily, we don't need rules for our vehicles for every scenario.

We're teaching the system to recognize when the world around it is behaving strangely. This might be a duck in the middle of the road, or a car making a right turn from the center lane, or a bicycle blowing through a red light. The self-driving car's sensors give it a 360-degree view out to two football fields, allowing it to see more and react.

Google's work with self-driving cars has been promising so far. But what obstacles remain? We don't believe there are any insurmountable technical hurdles in our path. It is largely a matter of fine-tuning the system. Every day, our test cars are out on the streets, navigating the roads and learning the pitfalls of driving, then applying that data to become safer drivers. Our goal is to create a fully self-driving car that is at least as safe as a human driver. When we reach that point, we believe the technology will be welcomed into communities. We'll begin to transform mobility and make the world's roads a safer place.

ROBOTS AND MANIPULATION: THE NEXT FRONTIER IN MOBILITY

Manish Kothari

MANISH KOTHARI is President of SRI Ventures and is a Vice President of SRI International where he leads the creation of high-value venture opportunities. Manish joined SRI in 2013 as a business development consultant and entrepreneur-in-residence. He became a program director in the Robotics Program in 2013. In 2014, he moved to SRI Ventures as director of commercial ventures and licensing, with an emphasis on health care, engineering, and physical sciences. Prior to joining SRI, Manish co-founded and was CEO of Mytrus, which offers a cloud-based software as a service (SaaS) platform for direct-to-participant clinical trials. Earlier, he was Vice President of R&D quality and operations for Simpirica Spine. There, he successfully built a spinal implant for lower back pain and spondylolisthesis. He holds multiple patents and is the author or co-author of several peer-reviewed publications and book chapters. Manish received his MS degree and PhD in biomechanical engineering from Cornell University, and he was a post-doctoral fellow at the University of California, San Francisco. His bachelor of technology degree in aerospace engineering (summa cum laude) is from the Indian Institute of Technology.

Mobility has already transformed the way we live by expanding and untethering our relationship with technology, and I believe the next natural evolution is in manipulation. Our current world is divided between digital and physical, but with robotics, artificial intelligence, and manipulation technology, those two worlds are coming together in a way that will change how we interact with devices, and how they interact with us.

There is a clear benefit to using robotics in situations that would be dangerous for a human being. Tasks that immediately come to mind are searching a road for explosive devices or cleaning up an area after a nuclear accident. DARPA has invested millions of dollars in research for the development of artificial intelligence that can be used in warfare and disaster zones. It is critical that these robots be able to navigate and respond to unstructured environments because an environment in these situations will rarely be static. Developing devices capable of manipulating the world around them has the potential to save lives, but it can also have a massive impact on how we live day to day.

My journey in manipulation started back in 2007 with telemedicine. The company I worked for had installed physicians licensed in all fifty states in an office in San Francisco and was practicing telemedicine all across the country. At the time, remotely interacting with patients was challenging because we had issues with equipment—for example, malfunctioning wireless routers or patients not knowing how to use a laptop. By 2010, telemedicine became easier because we introduced tablets to the process. Tablets proved to be more user-friendly with patients and experienced fewer complications than computers. By 2012, we had wireless devices connecting to the tablets, making the system even simpler.

Gaining FDA approval for our devices was hard, but the most frustrating part of all was that we still needed to send a nurse to our patients' houses every three to six months. Why? To touch, poke, and prod them. We needed to learn and understand what was wrong with them, but it was also important that we interact with them on a much more personal level. Sending a nurse to the patient allowed us to know how they felt, to joke with them, and to understand their emotions—all important parts of science that could not be done by any mobile device.

The nurse's role in my telemedicine project caused me to ask several questions, the most important of which was this: why was the nurse successful? She was able to accomplish her tasks not just because she could manipulate, but because she could interact as well. She could understand emotions, which enabled her to joke with the patient. She could put them at ease and reassure them—important components of the patient experience that we don't want to lose. Fortunately, these abilities are coming to a head in our mobile devices. Our AIs can now personalize the experience an individual has with technology. When you combine manipulation and physicality with delight and personalization, I believe you have the next step in mobility.

In considering everyday, real-life applications, you might

imagine a man watching a football game on television. His mobile phone rings across the room, but when he gets up to answer the call, he misses his team's winning goal. Now we already have solutions to this problem in motion, such as a smart watch that would allow him to answer the call, but what if he wanted a beer? I believe that the next solution will be household robots that the consumer can control with a mobile device or through voice activation. Most importantly, robots will have an active physical presence in their environments.

Seeing this necessity for face-to-face interaction, which was unachievable through technology at the time to the extent required, is what led me on a quest to understand how I could bring that physicality into the home. I joined SRI, a nonprofit, independent research center that I believed would be able to take manipulation into the consumer market. SRI is responsible for some incredible projects, such as Apple's artificial intelligence system, Siri. Several prominent technology companies originated at SRI too, such as Nuance Communications, a speech and imaging services corporation, and Intuitive Surgical, a robotic surgery manufacturer that grew to become the world's largest robotics company. The amalgamation of all these projects is what attracted me to SRI in my pursuit of bringing manipulation directly into the home.

Robots obviously aren't new to the tech world. SRI, at the time called the Stanford Research Institute, built Shakey the Robot in the 1970s. A mobile general-purpose robot, Shakey could flip light switches, climb flights of stairs, and open doors, proving that, even decades ago, robots were capable of basic functions. The real challenge now is getting their functionality to be at a human scale. Robots typically move much slower than a human and don't have the same emotionality. Historically, they have also been expensive, but these things are changing, largely because of the revolution in mobile. Communications, sensing, perception, and processing are now easy to do straight from a mobile device.

The cost of robotic technology is dropping dramatically, to the point where you can have a robot in your home for under one thousand dollars. Now is the time to increase functionality and improve manipulation so robots can serve a purpose in homes. Back in 2007, the Stanford University Personal Robotics Program introduced the PR1 robot. It was programmed to perform simple household tasks such as moving objects and cleaning the floor. On paper it sounds great, but it had several major flaws. One, it was only able to pick up objects of a particular shape that happened to be in a particular position. Two, it moved at approximately one-eighth the speed of a human while performing its tasks. Three, the robot was telemanipulated by a human with a controller rather than functioning autonomously.

Reliant on human input and rigidly restricted are not qualities we want in household robots. Our kids certainly aren't going to leave things strewn across the floor in a neat and orderly fashion for the robot to pick up, so a device with these limitations isn't practical. These shortcomings do, however, present the opportunity to improve upon past technologies and make a fully functional robot.

Much of the technology enabling robotics progress originated from the mobile world, meaning we already have many of the required hardware elements available in our toolbox. For example, the sensors on the fingers of a robot prototype are the same as the capacitance sensors on a cell phone. These components are already widely used, so they're cheap to mass manufacture and are readily available. When you combine the increasingly precise sensors with mobile technology, we have the elements to build a robotic arm that is capable of such intricate motions as picking up a balloon without compressing it.

The beauty of sensors is that they present a wide range of functionality in terms of how the robot interacts with its environment. The same robot that used sensors to manipulate delicate objects is also able to crush a can in its hand. It can manipulate small objects and perform simple tasks—for example, picking batteries off the floor and placing them in a flashlight. We can build such a device for under five hundred dollars. Thanks to the legacy of the ongoing mobile revolution and the dropping cost of components, autonomous and affordable household robots are finally within reach.

If built today, the robot described above would likely be teleoperated, but the process to convert teleoperation to autonomous operation is already being developed. Robotic perception had been a problem in the past, but now we have artificial intelligence that can recognize a coffee mug without being told what the object is by a human. The system can go beyond merely identifying objects and can use perception in more complex situations, such as recognizing that a crumpled bath towel is a washable piece of cloth and putting it in the laundry bin with the other linens.

Activities that were once tedious are becoming automated. There is software that uses simultaneous localization and mapping (SLAM) to create a 3-D representation of a house so people no longer need to map their architecture manually. We have robotic vacuums that can navigate your living room as well as telepresence that allows us to work anywhere in the world, and mobility will continue to drive the next generation of technology.

Autonomous systems are possible now, but we need the manipulation ability to complement the artificial intelligence. Once those two components are working in harmony, we'll be able to create household robots with more complex functions, such as opening doors, cleaning shelves, loading dishes into the dishwasher, and more. The possibilities are innumerable, but robots can't be purely utilitarian.

I realized from my work in telemedicine that the consumer experience must not be focused solely on function, but also delight. Consider an example in which a robot is designed to interact with children. It reads them books before bedtime and plays games, but to be fully effective, the robot needs to connect with the children on an emotional level. It's important that the robot understands whether they are happy or unhappy and reacts accordingly. It needs to be infused with a personality and be entertaining. Fortunately, these are problems that can be solved through programming and mobile processing.

Beyond providing delightful experiences, the purpose of household robots will be to give support during simple moments of need. I was in Japan visiting an assisted-living home where there was an obvious problem. The facility had enough staff members to hand out pills and give injections, but there were not enough people to feed the residents. It was an incredibly difficult situation for the staff because it takes half an hour to feed a single person, which, therefore, must be done with some degree of efficiency. At the same time, it is a moment of extreme loss of dignity for the residents. People do not want to be fed, so the situation must be handled with grace and respect.

Technology that can replace a human assistant in this scenario helps alleviate some of the loss of dignity. People generally feel less vulnerable and more self-sufficient when they're able to use a tool to support themselves in a task, rather than having to rely on another person. Using robotics would also solve the capacity problem, freeing up human staff members for other tasks. We've already made progress in this area with inventions such as spoons that help stroke patients feed themselves. Where we are still lacking is in helping the people who aren't even able to lift the spoon, and that is where robotics will step in.

Improving quality of life is the driving goal with mobile devices, with delight and dignity being two key features to keep in mind as the mobile revolution progresses. Advancements are made every day, and the hope is that Rosie from *The Jetsons*, a multifunctional household robot, will show up soon. I believe that it will, and that in the near future we'll see such a device that costs under one thousand dollars in every home. Household robots probably won't be able to do everything in the beginning, but, personally, I hope it can at least water my plants. Little acts of assistance like that can make a big difference in your life and are solid points on which to build.

One day, when your child asks her robot to get her a beer, it will turn to her and say, "Sure, I will in eight years, 36 days and four hours." This is the power of artificial intelligence. It will be able to bring both delight and dignity, playing chess with your kids one moment and then tending to your elderly parents another. I think manipulation can provide a level of support that every household will be able to use in some way, and the versatility is what will truly make this technology so incredible. We are all physical creatures, and once the next evolution of mobility bridges the gap with manipulation, technology will be able fully to join us in the physical world.

THE CASE FOR AUTONOMOUS TRANSPORTATION

Paola Santana

PAOLA SANTANA is an entrepreneur, lawyer, and public policy expert. She is Co-founder and Head of Network Operations at Matternet, a Silicon Valley start-up creating the world's next generation transportation system using networks of flying vehicles. A Fulbright scholar, GoodxGlobal Tech Fellow, Gifted Citizen awardee, and graduate from George Washington, Georgetown, and Singularity University, she aims to enact comprehensive regulatory frameworks and commercial ecosystems allowing the first networks of flying vehicles for transportation in the world. Previously with the World Bank, the Organisation for Economic Co-operation and Development (OECD), and the Dominican Republic's National Elections Court and Constitutional Court, she has developed striking public infrastructure projects and designed strategic plans to integrate advanced exponential technologies into e-Government platforms. Paola enjoys exploration, disruptive thinking, philosophy, and the arts, and she dreams of advancing the human race forward.

Out of all of the incredible technological innovations being created across the globe today, I believe that autonomous transportation is among the most exciting. In Silicon Valley, we're creating the next paradigm of transportation that will dramatically change the way we access the places and things around us.

For as long as I can remember, I've had a drive to bring massive positive change to the world. My home country is the Dominican Republic, which is a nation of resourceful people but has a history of poor management and administration. I thought that I could bring about change in my country and the world through politics, so I became a lawyer. I won a Fulbright scholarship and came to Washington, DC, to study political systems and the future of the government. Then, I hit a wall.

I had followed my ambitions, earned an expensive scholarship, and landed in the political capital of the world, but once I was there, I couldn't see a future for political systems. Our governments, the institutions in charge of creating equal access to opportunities and of improving the quality of life for its people, didn't seem to be carrying out the main task they were conceived for, and I couldn't find a way to fix that, or even work within a broken system. I felt that one more Band-Aid to politics wasn't going to be enough to create the change that our societies need in order to work for all. So, imagine that you aren't from the United States and you come to what is a brand new country full of bright opportunities, and then you feel lost; you feel like you can't affect the way that things work. That's where I found myself.

Discouraged by that reality, I was about to give up when I realized that there had to be a better way. Maybe the political-legal box I had built for myself wasn't the best way to enact change. I left Washington, DC, and came to Silicon Valley, where I discovered a new system: technology. Since the political route had not worked, I decided to figure out whether technology could be a better, faster, and more impactful way to bring the dramatic, positive change I had envisioned for the world.

One billion people around the globe have no access to reliable roads. These are individuals and families who are physically disconnected from the nearest local clinics, markets, and schools. This is a situation that people from my home country deal with every day, and one that has a profound impact in the daily lives of one-seventh of the world's population.

That's what I witnessed when I visited Papua New Guinea. People there face countless obstacles to carry out daily social and economic activities due to the lack of transportation options. Every time a family member fell ill and the closest health facility did not provide diagnostic services or didn't have the necessary medication, someone had to walk about sixty kilometers just to get access to basic health care. On the occasion that there was someone who could offer a ride, it was not unusual for the vehicle to get stuck in the mud for half a day at a time. It is such a common issue that there are tons of workers whose only job is to dislodge trucks that get stuck on the road. People back in the Dominican Republic also deal with this situation every day, so it is a problem that hits particularly close to home for me.

I thought there had to be a better way for these people to live, to be able to have seamless access to their closest hubs of services and local commerce, and to allow them to create sustainable incomes and a path out of poverty. So together with my team, I began to analyze more closely the transportation paradigm.

Transportation is a problem in the developing world, but it also negatively impacts first-world countries, albeit in a different form. Four billion people today live in congested cities and megacities, where a large part of their day is spent sitting in traffic. Our ecological footprint is massive, resulting in lost hours of productivity and wasted fuel. The question is, as the world population continues to grow and cities expand, is this situation scalable at all? If you look closely, you'll find that the overabundance of road infrastructure creates the same problems as its absence altogether.

Imagine if every emerging city in Haiti or Malawi turned into a New York City. Those cities are going to have the same congestion problems faced by the most densely populated urban areas today. We're going to have cities that are unlivable and where the inhabitants are trapped in traffic most of the time. Before the infrastructure gets to that point, do we have another solution? Have we rethought the way we are building roads, moving from one place to another, and sharing our public spaces?

Those are the questions my team was asking when we first approached the problem. The solution we devised is a network that functions like the Internet, in that it is decentralized, bottom-up, peer-to-peer, and works nonstop. The system is comprised of small drones, stations on the ground, and smart routing software, all of which are seamlessly integrated to form an ultraflexible and decentralized logistics and distribution network.

We call this system the Matternet—an Internet of matter. For the end user, it is a similar experience to using a smartphone app. As an example, the user can open an app, input the destination of the nearest pharmacy, and add their requirements. With a couple of taps in the app, that user has created a virtual, aerial route connecting her house with that pharmacy, and with one more tap she can see a drone taking off, flying autonomously to the pharmacy, and retrieving her medicine. On our end, the challenge is in building the backbone of that system to ensure it is reliable, safe, and sufficiently autonomous. The delivery vehicle, however, already exists.

The smart drone can carry up to one kilogram (2.2 pounds) over twenty kilometers (approximately twelve miles). It is a completely autonomous device that uses GPS and sensors to determine its location and destination. Flying at an average height of four hundred feet, the smart drone doesn't conflict with any other airspace. They fly to and from landing stations, which provide a demarcated space for the drones to approach ground level with precision, meaning they won't land in someone's pool, be torn out of the sky, or crash into anything. They will only land in their designated areas, which in the future can also function as ground stations where the vehicles can exchange batteries and packages automatically, allowing them to fly further than twenty kilometers if needed.

The Matternet software infrastructure forms the brain of the system. Every time a user makes a request between one takeoff location and one landing location, the software maps the best route between those two points. To facilitate flight planning, we create virtual roads that only include areas where the drones are permitted to travel through. They won't fly over schools, parks, or any other sensitive areas. When a request is sent to the cloud, the algorithm considers both speed and safety to determine an optimized path. The software is as much a breakthrough as the drone technology itself, and they both rely on one another to function.

For years now, many companies have created drones with cameras on them that are essentially cool toys. They have some practical uses—aerial maps, data gathering, and surveillance—but now we're taking the technology so much further. The ultimate goal is for the drone to become a physical transport vehicle for goods, which we believe could be the biggest advance in transportation technology since the invention of the internal combustion engine. For the first time in history, we have a vehicle that is mainly software and very little hardware. At less than one square meter, the drone is small and energy efficient. When you have a vehicle that weighs only four kilograms and can carry just under its own weight, the energy cost is relatively low compared to larger vehicles.

Most other lightweight vehicles still need the driver, which means you need a lot of supportive mass to move several pounds of cargo. If you're transporting a smaller amount of weight, you can use a drone to fly from point A to point B without the bulky infrastructure, requiring only a fraction of the energy needed for a driver-controlled vehicle.

Considering how much more convenient, cost-effective, and safe using a drone for autonomous transportation is compared to conventional methods, I believe this technology will totally disrupt the way we move and get access to goods around us.

Autonomous transportation should particularly impact the logistics industry. The logistics industry is an 800 billion dollar industry, with 400 billion dollars coming from the last mile that a product travels to its destination. More than 50 percent of goods we move are less than one kilogram, with a whopping 86 percent of goods weighing less than two kilograms. It is easy to imagine how we'll revolutionize the way products are delivered in the first and last mile once drone technology is fully matured.

As a lawyer, I know all too well that there are a number of public and regulatory challenges we must first overcome. Certain problems we can predict, while others may take us by surprise. Of the problems we're already anticipating, public perception of autonomous technology is an important one. As individuals, we each decide on a daily basis whether to trust another person or not, but we don't necessarily trust robots. This is a fact. We have instincts when assessing a human for trustworthiness that simply don't apply to artificial intelligence and technology. It's new territory with unknown potential dangers, and people don't know what to expect.

How can we take this technology to the point of reliability where we know that, 100 percent of the time, it will not make mistakes or cause harm? How do we solve the inherent problems? The first step is to build a very high-quality piece of technology that goes beyond being a toy. Once we have reliable technology, we need to inform the public of what to expect, how to interact with it, and have societies realize that this new era of autonomous transportation might start with small drones. We believe that in the future, with millions of drones flying in the sky above us, we need to establish mechanisms to insure and protect flight operations in the air and the people and property on the ground.

The second step is to enable aviation authorities to define the conditions under which they will integrate these flight operations into their national airspace system. This is an important problem to tackle because no aviation authority has a perfect formula or precedent on successfully integrating millions of drones in the current airspace, nor do they know how to mitigate all the risks that these low-altitude operations pose. The scope of the issue is vast, but authorities are beginning to propose rules using risk-based approaches focused on the overall risk of the flights given the type of drones, the complexity of operations, and the expertise of the operator, as opposed to just considering one of these elements or to addressing them separately. Opinions relating to potential rules are varied, with many insightful questions being asked regarding issues such as in what type of airspace the drones should fly, at what altitude, how to deconflict their paths considering their small size, and how to oversee a whole new system that we didn't anticipate in the near future. People from both sides—innovation and regulation—are working to find technical, operational, and legal answers to these questions.

Over time, we need to change the public's perception of drones. Currently, people equate drones with military technology. People in the Middle East, who may know someone who was killed in a drone strike, certainly don't want more of this technology in their countries. Even in the United States, we have negative associations with drones bombarding countries halfway across the world, or we might worry that someone could fly a camera up to our bedroom window and invade our privacy. The problem is that, after a piece of technology has been put to bad use, you cannot erase its history. While we can't change the drone's past purposes, we can influence its future.

If you take the technology from the hands of people who are not using it properly and put it to good use, you can make it life-giving instead of life-taking. I wake up early every day and go to bed late at night because I know we can do good with this technology. I saw the benefits, firsthand, that autonomous transportation can bring to people when my team was in Bhutan trialing the delivery of supplies from rural clinics to hospital labs, by request of the World Health Organization and Bhutan's Prime Minister. With only 700,000 people, Bhutan is a small, mountainous country facing the challenges of connecting rural health facilities or anything in the country seamlessly with other locations. We used our drones to fly from the national general hospital to clinics that were too hard to reach by road. It didn't matter that there was no good ground infrastructure built, because we could simply fly right over the terrain. Drones can facilitate everyday transportation needs, but they can also provide critical support in emergency situations when infrastructure is temporarily broken or when there's a need that surpasses the system's capacities.

One of the organizations that is constantly pushing the boundaries of what they can do in order to fulfill their mission is Doctors Without Borders. They go wherever they're needed, and literally, there are no borders or limits to what they can accomplish once they've identified a population that needs their help. This fierceness for solving humanitarian challenges has placed them among the top humanitarian organizations in the world and earned them numerous awards including a Nobel Peace Prize in recognition of their pioneering work in more than eighty countries. They inspire our team to be better and do more, and luckily, we had the opportunity to work together for the first time in 2014, after the government of Papua New Guinea requested their support to contain a tuberculosis epidemic in the country.

Doctors Without Borders sent people there to help with diagnostics and deliver treatment to patients. However, they were hindered by the fact that Papua New Guinea is 82 percent rural, with no way to enter the highlands. The organization called my team and asked us to test the feasibility of establishing a drone transportation network to collect tuberculosis diagnostics and deliver treatment to affected patients in an effective, reliable, and uninterrupted manner.

We partnered and set up a system wherein a drone would travel from a remote clinic to the main hospital carrying diagnostic samples, to be then retrieved and analyzed by hospital staff. What took them four hours by car, traveling from clinic to hospital in very dangerous conditions, took the drone less than an hour. With our technology, we were able to bypass the problems that people face there every day on the roads and avoid the delay in transportation of critical health-care resources.

The experiences in Bhutan, Papua New Guinea, and more recently in Malawi working with UNICEF have been eye-opening and transformational: we've seen how technology can make a huge difference in the lives of simple, ordinary people, and create new opportunities for them to have access, to have hope, and to thrive. Not only did using drones get people real access to health care, but it also built good will. People understood that, when they saw one of our drones approaching, it was bringing something positive. The technology was unlike anything they normally interacted with, but the community acknowledged that different isn't necessarily bad; in fact, they started to think that different and new was amazing, and that this could open up opportunities that they had not envisioned before. I think this proves that, given time and authentic interest in observing and understanding a problem and hard work, public attitudes can change, we can build high-quality technology that is reliable, and we can begin to trust autonomous transportation technology.

Being a lawyer and trying to bring about change through law and policy didn't work for me, but when I suspected that change would be possible through technology, I made a bet and embarked on a new adventure that allowed me to discover a world of rapid change, direct status quo defiance, and unlimited possibilities. I believe that there is no way to more directly connect people with the goods that surround them, whether they be diagnostic tests, presents for loved ones, or lifesaving medicine, than this groundbreaking new technology. There is still work to be done, but we're approaching the point where we can ask the question: what will we put in these payload boxes that will make a difference in someone's life? What will the transportation network of the future look like, where will it emerge first, and what will it start carrying? And I am honored and excited to know that it's up to us to set the course of change for how the whole world connects, as we create this next paradigm of transportation.

PIONEERING THE LUNAR FRONTIER

Bob Richards

BOB RICHARDS is a space entrepreneur and futurist. He is a Co-founder of the International Space University, Singularity University, SEDS, the Space Generation Foundation and Moon Express, Inc. a lunar resources company competing in the $30M Google Lunar X PRIZE, where he currently serves as President and CEO. Bob chairs the space commerce committee of the Commercial Spaceflight Federation and is a member of the International Institute of Space Law. As Director of the Optech Space Division from 2002-2009, Bob led the company's technology into orbit in 2004 and to the surface of Mars in 2007 aboard the NASA Phoenix Lander, making the first discovery of falling Martian snow. Bob studied aerospace and industrial engineering at Ryerson University; physics and astronomy at the University of Toronto; and space science at Cornell University where he became special assistant to Carl Sagan. Bob is an evangelist of the "NewSpace" movement and has been a catalyst for a number of commercial space ventures. He is the recipient of the K.E. Tsiolkovski Medal (Russia, 1995), the Space Frontier "Vision to Reality" Award (USA, 1994), the Arthur C. Clarke Commendation (Sri Lanka, 1990) and Aviation & Space Technology Laurel (USA, 1988). He is a contributing author of "Blueprint for Space" (Smithsonian Institution 1992); "Return to the Moon" (Apogee Books 2005) and "The Farthest Shore" (ISU Press 2009). In 2005 Bob received a Doctorate of Space Achievement (honoris causa) from the International Space University for "distinguished accomplishments in support of humanity's exploration and use of space."

Like many others, I grew up during the era of space exploration. Some of my earliest childhood memories are of the Rocket Garden at the Kennedy Space Center. Even at a young age, I wanted to learn everything I could about space exploration. My family lived north of Toronto, and I remember sitting in the family living room and watching broadcasts from Buffalo, New York, on our little black-and-white television. I was enthralled by the ghostly images of these brave men bouncing around on the Moon and completely bought the promises that we'd live in a future full of flashy technology and space adventures.

I believed in this future, not just because I'd seen Neil Armstrong and Buzz Aldrin on the Moon, but because I saw space everywhere. I could be watching the Moon landing and change the channel to see Captain Kirk rocket through the galaxy in the Starship Enterprise, or I could go to the theater and watch *2001: A Space Odyssey*. I would dream about what it might be like to meet beings with such advanced intelligence that they appeared god-like, and I couldn't wait for this exciting future to arrive.

Then, I grew up in a world where that didn't happen, and so I was, unsurprisingly, disappointed. We didn't have a Moon base or our wheel in space. There was no Mars colony. All of the advancements in technology I'd dreamt of were taking much longer to develop than I ever considered as a child. I feel that the people of my generation are the Orphans of Apollo—given a promise of adventure and uncharted territories that wasn't fulfilled. Fortunately, as adults, we have the chance to take innovation into our own hands and make our dreams reality.

One of the ironies of the space age is that only by going out into space and looking back on Earth did we realize the fragility of our world. That realization led to advancements in conservation, protectionism, and the idea of limits to growth. It's true that we live in a world of finite resources where we fight over land, energy, and materials, but we also must realize that we're a small island in an ocean of space that has infinite resources. A wealth of untapped resources is one of the many reasons it's important that we invest in space exploration, and while many potential resources are still far out of reach, there is one that's accessible now.

One of the great realities of our planet is that we're not just an island alone in the ocean. We're actually a two-world system, and our sister world is the Moon. It's a repository with trillions of dollars of untapped resources. You can think of the Moon as an eighth continent separated by a little bit of ocean that we have to learn to navigate. Crossing the ocean of space is still a challenge, but we've made significant progress on navigating the Moon.

Since the mid-2000s, robotic probes have allowed us to gain a deep understanding of what's on the Moon. We have mapped out more of the Moon than we have the Earth and now know what materials we can find there. The reason there are so many precious resources on the Moon is the same reason we have valuable resources on Earth. Every material that we use here—platinum, gold, and silver, for example—originally came from outer space. All of the heavy metals were bombarded by asteroids in the early history of our solar system. The difference between the Earth and

Moon in this scenario is the Earth is a much larger object, which means it remained molten for longer and many of the materials that struck the Earth sank into its surface.

The molten condition of the Earth is why we have to drill into the crust to collect certain resources. The Moon was also struck by asteroids but cooled more rapidly than the Earth because of its relatively small size. When asteroids hit the Moon, they were shattered rather than buried. Every crater you see on the Moon is a scar from an ancient bombardment, and the result is that the material from the asteroids is mostly scattered across the Moon's surface. If we develop the technology to collect and transport materials from the Moon in any significant capacity, we'll have an abundance of resources to choose from.

The real game changer that was discovered recently is that there is water on the Moon. You might wonder how big of a deal that is—it's huge. Think of water as the oil of the solar system, with its hydrogen and oxygen atoms being rocket fuel. Water is a game changer not just in our pursuit to liberate the Moon's resources, but for exploring the entire solar system, because it means we have a fuel depot in the sky. We need only to learn how to harvest these deposits of water and turn them into rocket fuel and other means of survival. I believe that the companies able to do this first will make Exxon look like a lemonade stand. With resources out there waiting to be claimed, the Moonrush has already begun.

Founded in 2010, Moon Express is a robotic lunar exploration and transportation company that aims to tap into the Moon's resources for the benefit of life on Earth and humanity's future in space. We believe that Earth is an island starving for resources and that we have a supermarket just next door. Our goal is to disrupt space transportation beyond Earth's orbit and solve the problem of getting to the Moon. We're closer to success now than ever before, armed with exponential technology and computational power that is far superior to what NASA had when they landed the first humans on the Moon. Space is quickly becoming accessible to those not affiliated with government agencies. Entrepreneurial groups of people can now create the technologies and spacecrafts necessary to reach what was only attainable by superpowers before.

Moon Express is building a breakthrough robotic spacecraft system that can make it from Earth's orbit to the Moon to explore for resources. We're concentrating on a micro-miniaturized spacecraft called MX-1 that will serve as a secondary ride on a rocket. Instead of requiring a whole rocket just for the MX-1, we fit the craft, which is approximately the size of a large coffee table, as a hitchhiker underneath a primary payload. This method reduces the mission cost to a fraction of the average space launch, from nearly one hundred million dollars to ten million dollars. Relatively speaking, the MX-1 will be a cheap ride into space.

Once we're in space, we plan to function as a platform. The MX-1 isn't just a lunar lander—it's a spacecraft workhorse. The goal is to be the iPhone of space, in that we'll reach the Moon, but we'll have many other applications as well. Possible functions include planetary defense, Earth observation,

CubeSat deployment, satellite refueling, security operations, space debris collecting, and space tugging. Our maiden mission to the Moon will be a technology demonstrator, partnering with Google and NASA, to show that a private company can do what was once only possible for government entities. We think that mission will be a singularity moment in space, redefining what's possible and opening new markets to us with low-cost lunar access and rideshare opportunities.

Our second mission will be to deliver an observatory to the south pole of the Moon and set up the first Moon camp. People will be able to connect to the observatory through the Internet and look out into the universe, gaze back on Earth, and see vistas previously only witnessed by astronauts. Pay dirt will enter the picture as our third mission when we bring a sample back to Earth. We predict that a lunar sample the size of a basketball will be worth about one billion dollars. At that point, the authorities will have to approve the concept of gathering material from another world, bringing it back to Earth, and selling it, and they will need to initiate the case law necessary to say that that private sector can do business not just on Earth but on other worlds.

I feel confident that we'll reach our goals, largely because of the incredible minds we have working on this project. Moon Express co-founders Naveen Jain, Barney Pell, and I have pulled together a great executive team comprised of some of the most experienced engineers in the world, with hundreds of missions between them. We've also populated the company with what I like to call MOONhackers, young people in their twenties whose energy and willingness to take risks compliments the skill and wisdom of our more experienced team members.

Our strategy involves taking technology from NASA and applying Silicon Valley-type rapid innovation, prototyping, and testing to it. The team adheres to the mantra "fail early, fail often, fail digitally." Just across the road from our headquarters, we were able to invest in a spacecraft platform at NASA Ames called the Hover Test Vehicle. We found our training wheels with this vehicle, using it to our advantage for software development and testing.

After the Hover Test Vehicle, we graduated up to the Mighty Eagle, a vehicle at NASA's Marshall Spaceflight Center. We upped the game a bit and, keeping with this analogy, earned our driver's license. While it was NASA's hardware running the program, it was our own digital assets that made the vehicle fly. The Mighty Eagle uses hydrogen peroxide, which is decomposed over catalyst beds. This reaction provides the propulsion and is vented by the vehicle as high-powered steam. Our flight tests took the Mighty Eagle up to a height of about three hundred feet, which feels exceptionally high when you're dealing with another company's asset. Our goal was to land right-side up, and in doing so prove the validity of our guidance navigation and control software so we could then invest in our own vehicle.

I'm happy to say that the Mighty Eagle landed upright, so we were ready to build our own vehicle. We invested several million dollars into the project and hired Tim Pickens of SpaceShipOne fame. He was a SpaceShipOne engineering lead for their propulsion system and had built a rocket engine based on the same hydrogen peroxide technology we planned to use. Traditionally, every good Silicon Valley company starts in a garage, and we tested our first rocket engine in Tim's garage in Huntsville, Alabama. The hydrogen steam engine was a success. One of the reasons why it was important that we used hydrogen peroxide is because water, H_2O, can become H_2O_2. That chemical reaction means we have a fuel depot waiting for us on the Moon.

After developing our engine, we partnered with NASA under a program called Lunar Catalyst. In the same way that SpaceX and Orbital Sciences partnered with NASA to deliver cargo to the space station, NASA has partnered with us to eventually deliver cargo to the Moon. Using our engine, we built our first prototype spacecraft, went down to the Kennedy Space Center in Florida—back to the place where I had my earliest childhood memories—and flew our vehicle.

We are going back to the Moon, and we're going to fulfill that vision that we Orphans of Apollo had as children. There's a whole sky full of untapped resources, and once we reach those heights, our economic reality will change. We'll be a multiworld species, with a new version of trade between the continents. Only this time, it will be trade between worlds. I'm proud that Moon Express is one of the vanguard companies as we move into the next era of exploration. Look toward the sky as we launch out of Kennedy Space Center and take our mission one step closer to the Moon. It is my hope that someday, if you're so inclined, you too will be able to board a shuttle and join me as we pioneer the lunar frontier.

THE MOBILE FRONT LINES

Ezra Kucharz

EZRA KUCHARZ has more than twenty years of experience creating, evolving, and managing a number of leading Internet properties. He has served as President of CBS Local Digital Media since January 2010, and oversees the online and mobile strategy for CBS Television Stations and CBS RADIO, using the assets of more than two dozen television and 126 radio stations. During his tenure, Ezra has developed a vast platform of news, sports, and entertainment Web destinations, bringing together the most trusted brands in media and offering visitors a truly "local" experience. The division's digital offerings include live and on-demand audio and video streams, original programming, editorial features, and coverage of live concert events. Ezra earned his bachelor's degree from Boston University in biomedical engineering and worked for NASA in their space shuttle and space station medical operations departments following graduation. He later earned masters degrees in engineering management and medical informatics at the University of Houston and Duke University, respectively.

Today's mobile devices need certain pieces of hardware to function—power sources, touch-screen user interfaces, and processors—but they also rely on something less tangible than their metal and plastic parts: content. Content is the fuel that we consume every day of our lives, and without it, our mobile devices wouldn't be nearly as useful. Your choice of what to read, watch, and listen to are what turns an off-the-shelf mobile device into something that is uniquely yours. If you're skeptical of content's importance, I challenge you to put away your phone, television, books, and magazines for one day and see what happens. Most of us would start to go a little crazy or, at the very least, feel uninformed, isolated, and bored. Content is something the human spirit needs.

Mobile devices have started a new chapter in how we connect with one another, capture real-time events, and circulate information around the globe. The days of scratching our heads trying to remember the date of an event or the name of a catchy song are over. We now have the power to answer those questions in the palm of our hand. Mobile has expanded the pool of knowledge we're able to tap into on a nearly limitless scale, and untethered Internet access allows us to sate our desire for information whenever and wherever the need strikes. As a result, we're changing the way we develop and think about content.

For the media and other content creators, mobile has been both a blessing and a curse. We're free to reach people in new ways but challenged by the explosion of sources. Content distribution is no longer a privilege belonging solely to media power players but is available to everyone. At the turn of the twentieth century, there was no way for the average citizen to create and distribute independent content on any significant scale. A city might have had two or three newspapers serving as the primary channels for people to consume timely content. Behind those newspapers, a relatively small group of people decided what information would be made available to the public.

Several decades later, commercial radio was invented, which allowed people to tune in for live information, contributing another handful of content sources to the pool. By the time television was introduced to the mainstream, which added even more sources of news and entertainment, content had become a commodity. Up to this point, content distribution was still limited to media tycoons, but then technology changed everything. Today, with the Internet and mobile devices, consumers have their choice of countless content sources.

The tools needed to create content are more available and easier to use than ever before, and plenty of people make interesting content on their personal computers and mobile devices. Even for decades prior to the Internet, there was nothing stopping a band from going into their garage and recording great music. The possibility of creating a breakout record was there, but here's a little-known fact: in the past fifty years, there have only been two number-one hit songs by independent artists. Those artists were Lisa Loeb, who made it to the top in the 1990s and, more recently, Macklemore. Every other number-one song in that time frame has been produced through the record label system. The media companies are still coming out on top, but the game has changed.

Making exceptional content remains difficult, but when anybody with a device and Internet access can create a blog or post a video to YouTube, those of us who make our livelihoods in creating content are required to step up our game. We need to stand out from the crowd with consistently high-quality products and treat content creation like the skilled craft that it is. It's important to note that the type of content I'm talking about excludes two classifications: perishable and commoditized.

Perishable content is defined by its shelf life. It conveys timely information that consumers want to know as an event unfolds, after which the news ceases to matter. An example might be last week's sports report; fans want to know who won the game, but then the content expires. Perishable content is valuable and needed, but it doesn't engage or drive consumers in the ways that other content can. Commoditized content consists of widely available information that is difficult to present in a unique light. In addition to perishable information, commoditized content also lacks the ability to elevate its source and impact consumers. A good example of a commodity is the weather forecast. The information may be packaged differently, but it is essentially the same from all sources.

One way media professionals are differentiating themselves is by thinking about content less as a series of standalone articles or video clips and more as a multichannel experience. At CBS, we operate hundreds of Web sites, apps, television channels, and radio stations, and reach approximately forty million people each month on mobile products alone. Our focus is on integrating our different channels to reach consumers wherever they happen to be, and they happen to be everywhere.

An experience of mine opened my eyes to how mobile has changed the way people consume content. The Final Four basketball tournament was taking place in Indianapolis. We saw Duke (where I went to graduate school) beat Michigan, and then we left the second game at halftime to eat at a nearby steakhouse. In that restaurant, something magical happened that I had never seen before. More than two-thirds of the tables had a phone propped on its side, held up by salt shakers and butter dishes, watching March Madness on demand. Everybody was viewing the game on a mobile device.

The particular brand of the devices wasn't a distinguishing feature. Everybody in that restaurant had different phones—Windows OS, iOS, and Android OS—yet they were all digesting the same content. What mattered is that they were accessing the content through mobile. I've been working at traditional media companies for over fifteen years and had never witnessed this type of mass mobile content consumption before. Not so long ago, it wasn't even possible, but we're living in the mobile era. That moment reinforced everything we had been talking about at CBS regarding our content strategy. Mobile both allowed and necessitated an alteration to our approach. The answer was integrated content.

In my experience with mobile thus far, omni-channel content has outperformed pure-play. At CBS, we've had success integrating mobile into our live broadcasts, with the results being a wider and more engaged audience. One standout example is a documentary we made several years ago with Taylor Swift about her album-of-the-year nomination. The goal was to integrate live broadcast with mobile streaming, so we ran a teaser clip of the documentary on our television stations and directed people to watch the rest on mobile. The payoff was several million people watching the full documentary on their mobile devices. We took the same approach with the Grammys, capturing consumers' attention with pre-event interviews, behind-the-scenes videos, and trivia on mobile devices to drive them to the live broadcast.

Another example of cross-channel content involved baseball player Alex "A-Rod" Rodriguez. Alex was embroiled in an investigation into his possible use of performance-enhancing drugs. One day in New York City, he was seeing the arbiter about his case when he stormed out of the courthouse. Feeling that he'd been unfairly portrayed to the public, he immediately headed to the CBS radio station to give his side of the story. We interviewed him and published the clip to CBS's various channels. In the weeks that followed, the interview clip was listened to nearly five million times on mobile devices. Omni-channel distribution helped the content reach a large audience, as did the fact that it was a noncommoditized story unique to our station.

My final example of mobile as an effective content channel

is the Boston Marathon tragedy coverage. Emergencies of this scale are terrifying, and, naturally, people want to know what's going on as soon as possible. The Boston Marathon attack took place during the afternoon when people were generally out of their houses. CBS covered the story across our channels and had 2.9 million people watch video coverage on mobile and 2.1 million people listen to our radio stations on mobile. If we had only broadcast our coverage on television, millions of people would have been without information in a time of national panic. Having mobile access to live updates was even more critical for the citizens of Boston, who needed to be aware of potential dangers in their area. Mobile content is an effective way to provide entertainment and news, but I think this example demonstrates how critical a role it can play in serious situations as well.

Technology and the media have a symbiotic relationship that seems, at this point, inextricable. Our mobile devices require content to be useful, and content needs a way to be seen. The media industry will, out of necessity, have to change as technology continues to evolve, but among these changes are exciting possibilities. We can engage consumers on multiple fronts, provide a deeper content experience, and explore creative new ways to combine channels. The delivery system is primed, but to succeed in an arena where anyone can play, we need to earn our place on consumers' screens. When you think about the next shiny new device, remember the role content plays in technology's usefulness and the power it can wield. Content is king, and when you combine the vast reach of mobile with uncompromising quality, the result is worthy of the crown.

HOW MOBILE HAS CREATED A NEW WAY TO PAY

Aditya Khurjekar

ADITYA KHURJEKAR is CEO of Let's Talk Payments (LTP), a digital destination for global FinTech insights. LTP has been the fastest-growing source of curated content and data-driven research since 2013, and now reaches sixty thousand subscribers globally through its daily newsletter. Previously, Aditya was Executive Vice President at nD Bancgroup, a bank holding company that enabled mobile money experiences for consumer brands. In 2012, Aditya cofounded the breakout conference on emerging payments and financial services, Money2020, which now attracts more than thirteen thousand attendees every year. Until 2011, Aditya was the New Business Development Executive responsible for all mobile payments and commerce initiatives at Verizon Wireless. Since 2008, as a founding member, he had business ownership for Softcard, the mobile commerce joint venture between AT&T, T-Mobile, and Verizon. Aditya holds an MBA in Finance and Management from New York University and an MS in Electrical Engineering. He lives in Charlotte, North Carolina, and works with teams all over the world.

Throughout history, humans have relied on coins, paper, plastic, and other physical tokens to embody the value of goods and services. Carrying and exchanging physical money was the default system, but when the mobile revolution swept across the globe, the paradigm changed. Payment technology is a space that had been starved of innovation for a long time, but mobile payments have fundamentally altered the way we do business and are rapidly becoming the new standard.

When we think about mobile payments, the first action that comes to mind is making a payment through a smartphone app. However, the true definition is broader than that. As a mobile phone user, any transaction you make that is tied to your phone number or mobile service provider—whether that is making a payment, saving money, conducting commerce, or anything related to managing or moving funds—falls under the umbrella of mobile payments. The key to these transactions is that each user has a phone number that uniquely identifies him or her. Simply having that identifier solves one of the most significant challenges with payments, which is confirming that the person promising to pay you is who he says he is.

There are several areas in which mobile payments are substantially better than their predecessor, the plastic card. A mobile device is connected, so it can talk to the Internet and make use of external resources. It is interactive, in that the consumer can tap, type, or view data within an app. It is also intelligent and can make decisions, such as whether an inputted password is correct or not. This connectivity, interactivity, and intelligence are the primary attributes that give mobile devices the edge. When you bring those benefits into your payment experience, it is clear why mobile payments are superior to plastic.

Payment in its strict definition is just one part of the online shopping experience made easier by mobile. The rest of the comprehensive digital experience includes actions such as clicking a coupon, finding a deal, viewing ads, or filling a digital shopping cart. All of these experiences fit together and are facilitated by first establishing your mobile identity. Signing into an app to make a purchase is faster and more secure than handing somebody a credit card at the store, and the consumer is able to make a transaction without ever relinquishing control of his or her data.

Why are mobile payments more secure? Consider the certainty you feel when you make a phone call to someone. You know you will speak to the person you are trying to reach. It is not impossible to undermine this connection, but it is extremely difficult for someone to claim a fraudulent identity when making a phone call. A good point of comparison is sending a physical piece of mail. When you send a letter to someone, faking your identity is as simple as writing a different name and address on the envelope. There is no easy way for the receiver to confirm your identity.

Phone calls from a mobile device are not the same. The fact that a person must register a phone number means that her device is inherently tied to her identity. When we apply this fact to payment, it is obvious that conducting a transaction with a mobile device is a more reliable way to identify somebody. It provides reassurance to both the person paying and the entity being paid. Being digitally connected also makes fraud a significantly more challenging crime, whereas with credit cards or cash, fraud is as simple as someone slipping his or her hand into your wallet.

Mobile payment technology has come a long way over the past decade. I personally began working in the space in 2008, when I took the role of New Business Development Executive, Mobile Commerce and Payments at Verizon Wireless. At the time, my team and I were starting to look at mobile commerce from a mobile operator perspective. We wanted to find out how we could use the assets we had as a mobile company to add value to the payment-connected services industry.

Our primary challenges involved creating the hardware and software needed to help subscribers make mobile payments and transfer money. These challenges were not unique to Verizon Wireless, of course, but were issues that many companies had invested money in solving. Some of the mobile initiatives taking place across industries succeeded, while many others did not. For example, banks were trying to charge their customers continuous fees to use mobile banking. Today, if a bank told people to pay ten dollars every month for mobile services, the bank would be ridiculed.

Mobile payments are something that we've now come to expect. The service is an integral part of the proposition a bank might make, when offering a customer a checking account for example, and a lack of mobile services would be conspicuous. Consumers' expectations and the technology to back them up are both developing at unprecedented rates. If someone told you just a few years ago that, today, you'd be able to tap your phone at the McDonald's kiosk and pay for your burger, you probably wouldn't believe him. Yet, tap-to-pay services and digital wallets such as Apple Pay are now part of our reality and are securing a foothold in the marketplace. The technology isn't implemented by every company, nor is it always a seamless experience, but mobile payments are quickly gaining adoption.

Where we have already seen widespread mobile payment use is with apps such as Uber and HotelTonight. When you use one of these apps to book a car ride or hotel room, you're using your mobile signature to prove your identity and make a payment without having to enter a card number or password. Consumers use these apps every day without even thinking about it because, in many cases, it is actually easier to make a transaction on the phone than in a browser. Seeing the success of mobile-based platforms, companies are investing heavily in app development and growing their mobile presence.

One of the most recent features to be released in mobile payments is contactless technology, a capability that mobile carriers have been working on for the past ten years. Contactless payments require a special secure connection with the mobile phone similar to other specific wireless connections that are made possible by a variety of components in the device. These components include a radio communication chip for making phone calls, a Wi-Fi chip for connecting to the Internet, and a Bluetooth chip for syncing the phone with peripherals. As mobile phones advance, manufacturers add new radio frequency chips to facilitate different types of wireless communication. The newest chip type, near field communication (NFC) technology, is what enables contactless payments. NFC technology has a secure, low-range communication protocol that allows payment information to be sent from the phone to the payment terminal when the two are in close proximity to one another.

A huge amount of collaboration was required in the tech industry to implement NFC chips and the apps that use the technology. Device manufacturers such as Samsung, LG, and Apple worked closely with operating system developers for iOS, Android, and Microsoft to ensure that the hardware was compatible with their platforms. In turn, the operating system teams collaborated with app developers to bring the technology to life.

Making contactless payment happen ended up being a long process of experimentation and fine-tuning. The first country to release the technology was Japan, followed by parts of Europe, and then the United States. Now, every new smartphone sold in the United States contains an NFC chip. The challenge at this point is forging the complex business relationships required to put the technology to use. The lack of those relationships is the primary reason that contactless payments haven't taken off as quickly as other mobile payments. There is both a technological aspect and a business side needed to bring NFC applications to fruition.

Another innovation that makes mobile payments superior is blockchain technology. Blockchain technology is a new way of structuring information that allows us to transfer money on a global scale in a much more secure manner than the old methods. Traditionally, it has been necessary for banks to invest in huge amounts of security technology. They often keep their data centers in undisclosed locations with a redundant infrastructure so that their customers' information is not compromised in transit. The drawback is that banks have to spend huge amounts of money on these measures, and that cost is passed on to the customer.

Blockchain technology disrupts the money movement process by distributing information over many different computers. As a result, a hacker would have to infiltrate all of the computers containing bits of the target information at the same time in order to obtain that data. Breaking the system becomes practically impossible, and the cost of security for banks and other companies is consequently much lower. I believe blockchain technology is among the top changes that will have the most profound impact on financial services over the next decade. It may not be a change that consumers will actually become aware of, but blockchain technology will redefine the infrastructure behind banking and commerce as we know it.

Blockchain could completely disrupt the way banks have approached the money-transfer business—through their correspondent networks that are much like the alliance networks that airlines have built their loyalty programs around. With billions of people, particularly those in emerging economies and in remote places, now having access to the Internet through their mobile phones, banks and money-transfer services can use the blockchain to move money between any two parties, anywhere in the world.

Not only is payment technology changing, but the intention behind our buying behaviors has shifted as well. As a consumer, you don't wake up in the morning wanting to make a payment. Paying is a necessity, but given the choice, you would rather not. When you decide to purchase a pair of pants or a car, it is an example of intent-driven payment. You make a conscious decision to spend money on those items. With apps and mobile services, we are now entering the realm of event-driven payments, which are much more passive.

Event-driven payments are triggered by an experience or service, such as taking an Uber ride. When you request a driver, you've already installed the app and done the setup, so you never have to take the discrete action of paying. Everyone has had the experience of taking a cab ride only to have the driver bump up the price or tell you he or she only accepts cash. Automatic event-based payments solve these problems and allow consumers to have a better experience than with traditional exchanges. When people are offered a new approach to a service such as Uber, where payment is straightforward and painless, they embrace it. This is possibly the best example of mobile payments saving time. Customers no longer have to sit idly by on a busy street waiting for a driver to make change or process your credit card; it's automatic, simple, and efficient.

Giving consumers the option to have a more enjoyable experience with mobile payment is putting pressure on companies to adopt the technology. Consumers will not necessarily spend more money because of a positive user experience, but that experience will dictate where they choose to take their business. I can say from personal experience that when I have to book a hotel room at the last minute, I'll check two or three apps for the best deal. However, the app that allows me to pay with just my fingerprint touch ID instead of the old-fashioned sixteen-digit-input credit card payment is the app that I'm going to use to pay. At the end of the day, convenience wins.

The risks associated with mobile payments are low, but worth examining. Mobile phone users are increasingly putting their information in the hands of app developers. We're so comfortable with using our phones to do things related to commerce and social sharing that we don't always think about what information we're giving away. For example, when you use Google or Facebook to log into a third-party app, you're making a huge decision to transfer your identity to that third party. Why do we do this without a second thought? The answer is convenience. We live in an age where convenience trumps trust, where a consumer provides companies of all shapes and sizes with access to personal information, payment information, and his or her broadest friend base. If we were actually to read the terms and conditions, we might think twice about what permissions we share.

This act of trusting a company with your information isn't

limited to third-party app developers, but applies to Google, popular social channels, and other large entities, too. We allow companies to know so much about us and to profit from that information because it is convenient for us. There is a fundamental consumer behavior shift toward trusting the mobile tools in our hands implicitly. People already value mobile photo sharing, mobile calendars, mobile email systems, and other services, and I believe that the mainstream will come to trust mobile payments in much the same way. My one word of caution is that people should be wary of how much personal information they give away for the sake of convenience. It is possible to enjoy the incredible benefits of mobile services while still being smart about protecting your identity, and we should all be mindful as technology continues to evolve.

More than ever, technology is playing a crucial role in innovative financial services and has improved the way we conduct commerce. The transition from coins and plastic to mobile is already well in motion, and the companies that come out on top will be the ones that embrace change. On the consumer end, mobile has made transactions so seamless that we don't even need to think about payment—we can simply enjoy the services we choose to use. All of this is possible because of the mobile revolution. The smartphone has changed financial services for the better, and the day is coming soon when in order to make most purchases, instead of pulling out your wallet, you'll reach for your phone.

VIRTUAL REALITY: THE FINAL PLATFORM

Bertrand Nepveu

BERTRAND NEPVEU has spent his entire career working in the field of computer vision. From a very young age, he knew that he wanted to be an entrepreneur like his grand-father and change the world. Hardcore gamer since Donkey Kong on the ColecoVision, he became obsessed with virtual reality when he tried the Power Glove on the Nintendo Entertainment System. As an early adopter and gadget lover, he founded Vrvana in 2005 in order to fulfill his dream of improving the gaming experience by bringing the user inside the game. Geek at heart, he gathered the best and brightest in Montreal to create the ultimate mixed-reality headset. Bertrand holds a Computer Engineering degree from the Université de Sherbrooke and an MBA from HEC Montréal.

The mobile revolution has reshaped our world, both physically and digitally. It unhitched computing from bulky machines and allowed us to communicate and consume information with a newfound freedom. When we look back at the history of computing, there have been several major platforms, each progressively more mobile than its predecessor. The smartphone has been the reigning platform since it gained ubiquity, but what's next? The answer is virtual reality. I believe that virtual reality will not only be the next big revolution in technology, but it will be the final computing platform.

My personal interest in virtual reality began a long time ago. I've been a hardcore gamer for most of my life, going all the way back to playing Donkey Kong on the ColecoVision gaming console in the 1980s. In the early 1990s, Nintendo released a game controller called the Power Glove that used sensors to detect finger movements. I remember trying the device at a store and thinking that it was a good first step toward virtual reality. That taste of movement-based gaming helped prompt me to imagine what the ultimate gaming experience would be, and I thought the answer was a virtual-reality headset. Being able to feel as if you were in the game rather than simply looking at it through a two-dimensional screen would take the gaming experience to a whole new level.

Considering that the concept of virtual reality has been around for decades, why are we suddenly seeing rapid progress in this technology? One primary reason is that, until recently, display technology was not adequate for this use.

For example, if you try to use a liquid crystal display (LCD) in virtual reality, you might have a passable experience, so long as you don't move much. You'll feel immersed in the virtual environment, but if you start turning your head, the details will blur. The LCD technology isn't fast enough to light pixels for only three milliseconds at a time, which is the speed required to smoothly display movement.

The only technology that can achieve those speeds at the moment is the organic light-emitting diode (OLED), which is relatively new. OLED displays have the necessary resolution and frame rate to provide an excellent virtual-reality experience when combined with improved tracking technology. OLED displays are now widespread enough in smartphones and other consumer products that virtual reality can reach the mainstream, which is why we're seeing an explosion of new development in this area.

In addition to displays, other hardware advancements were needed to bring virtual reality to the consumer market. One of the most critical components is the inertial motion unit of a device, which is comprised of the gyroscope, accelerometer, and magnetometer. These sensors have become faster, more affordable, and more precise in recent years, so they can accurately track a user's head rotation in virtual reality. Certain headsets also use an outside camera and sensors to track where the user is located in relation to his or her physical surroundings.

At Vrvana, our flagship headset, Totem, is capable of both virtual and augmented reality. The headset uses onboard sensors with cameras, including one that captures infrared light, to map the room you're in and measure in real time how you have moved in relation to the objects around you. The technology we're using is cutting-edge, and the features our headset has would not have been possible a few years ago.

We're now at a point where all of the technological puzzle pieces are available, so it's no coincidence that we're entering the golden age of virtual reality. Not only is the hardware affordable and fast, but we also have excellent game engines and a wealth of content at our fingertips. Even better, the barrier to entry for creating new content is low. To create a 360-degree video, anyone can combine six GoPro cameras and capture footage. We have a viable ecosystem where there is a constant influx of new content and huge potential for commercialization, which is why big studios are making investments in the space: they know it's the future of media.

While it's unfortunate that it took many more years than I anticipated for the tech industry to achieve a workable virtual-reality headset, it has been worth the wait. Once you're in virtual reality, you don't have any limitations. While I personally am thrilled to be able to play games with a headset, the applications go far beyond entertainment. Virtual reality could redefine our working environments. For example, we currently see people arrange two or three computer monitors across their desks, but in virtual reality, you could have an infinite number of screens. You could have virtual meeting rooms or even entire buildings, reducing the need for workers to commute to the office. People could have the ability to work from anywhere in the world without losing that critical face-to-face component. The incredible power of virtual reality is that you're not bound by physical limitations and can create anything you imagine.

The education industry is another space that could be revolutionized by integrating virtual technology. Online college courses are growing in popularity, but imagine if you could attend class in a virtual lecture hall, complete with the professor standing at the front using a whiteboard. You could recreate the immersive feeling of attending a physical classroom with added benefits. For example, during a lecture, a three-dimensional animation could appear in front of you that would explain the laws of gravity, quantum physics, or other complex concepts that are not easy to understand.

We could dramatically help struggling learners to succeed in the classroom and receive a high-quality education. Not only would the experience be more convenient than an on-campus class, and without the inherent distractions of being surrounded by other people, but every student could watch the lecture from the front row. They could move through lessons at their own pace in a much more interactive environment than a standard classroom allows. In this way, virtual reality in education will help to personalize lessons for an individual's unique needs.

In the near future, we can also anticipate the rise of augmented reality. Unlike virtual reality, in which you are entirely immersed in a virtual environment, augmented reality adds a layer of information to the real world. It will have powerful applications in enterprise and productivity.

For example, instead of having a small GPS device in your car, you would wear an augmented reality headset that would put directions in your field of vision while you drive. Drivers wouldn't have to move their attention from the road as they have to with a dashboard-mounted GPS device. In this way, integrating augmented reality into the driving experience could result in fewer collisions and make the road a safer place.

I'm convinced that virtual reality—and augmented reality, in its own role—will be the final platform. Once we master the technology, we can create whatever we want. We're close to seeing virtual reality's full potential and have made significant progress since the Power Glove, but there is still work to be done. The main hurdles are creating an effective user interface, developing the mechanics of virtual interactions, and producing affordable devices. For high-end experiences, price is currently a barrier, but consumers can already get a fairly compelling virtual-reality experience for under one hundred dollars. The drawback is that the most affordable offerings lack finesse and are reliant on the presence of a smartphone.

At the low end of the market is Google Cardboard, Google's first venture into virtual-reality headsets, which offers an introduction to virtual reality at a less-than-twenty-dollar price point. Cardboard is, as it's name implies, a viewer that is made out of cardboard and that holds a smartphone behind a pair of low-cost plastic lenses. The experience may not be as interactive as consumers hope for, but it provides satisfying 360-degree video playback, which will blow the minds of most people when they first try it.

In addition to Cardboard, there are mid-range solutions that also use the smartphone as a screen, but with better controls, better display, and additional tracking sensors in the one-hundred-dollar range, such as Samsung's Gear VR (for Galaxy smartphones) launched in cooperation with Oculus.

At the high end, premium virtual reality experiences cannot be powered by a smartphone. Oculus, HTC Vive, and Sony Playstation VR are self-contained headsets with their own screens but are powered by high-performance PCs or special gaming consoles. These products will have initial prices in the $850 to $2,000 range when you take into account the price of the console or PC.

I believe low-end headsets will function as an infection vector in that, for a low price, they offer people a taste of virtual reality. Those initial users will show their friends and family, and then they'll want a deeper experience. Eventually, they will commit to a fully interactive device at a higher price point. Furthermore, as we see the cost of computers and virtual-reality hardware go down, we should see a rise in the mass-market adoption of mid- to high-tier headsets.

In its execution, virtual reality is simpler than augmented reality because there are no external stimuli to take into account. Similar to how an architect designs a building, you can create the virtual world and determine its every aspect. A virtual-reality headset will need to render every pixel of the environment on a continuous and motion-responsive basis.

On the other side, augmented reality requires that only a portion of a screen be rendered, making it less GPU demanding and more mobile friendly. However, augmented reality needs to take into account and interact with the many uncontrollable variables of the outside world, which presents a significant challenge.

Imagine that several people, each wearing an augmented reality headset, are sitting around a conference table and manipulating a virtual object. For the brain to think the object is real, the interaction needs to be as lifelike as possible. If there's a light source, the virtual representation needs to display shadows. If someone puts his or her hand in front of the object, the display needs to show the hand masking that part of the virtual object. Numerous other externalities need to be accounted for, making augmented reality a much more complicated challenge that will likely take several more years to solve.

One of the most important components of virtual reality that needs perfecting is the coherence between senses. Humans have five senses, and if our senses aren't in agreement with one another, the body reacts poorly. The response is a natural survival mechanism that tells the brain something is wrong. For example, if you feel your head moving but don't see your environment tracking at the same speed, you will feel nauseous. For that reason, the synchronization between movement and sight in virtual reality needs to be perfect for it not to cause that effect. As developers and creators, it is critical that we make sure our headsets can render images and sound fast enough. Whether or not we can trick the brain into believing what it sees will play a huge role in the user experience and will determine the adoption rate of virtual reality.

As sensitive as the brain is at perceiving minor incoherencies, it is also incredibly adaptable. In my personal experience, the first time I tried a virtual reality headset I felt nauseous after five minutes. I thought a device that caused such a negative reaction would never succeed. Then I tried the same headset two days later and had the best time of my life. Having gone through that previous experience, my brain was able to cope with the second exposure much better. Also impressive is that, when the brain is convinced of what it's seeing, the elicited emotions can feel incredibly real.

There is great potential for using emotion-driven content in marketing, entertainment, or to sway people to a particular cause. For example, the *New York Times* mailed a Google Cardboard headset to each of its subscribers along with a 360-degree documentary on Syria. I watched that documentary and can attest to the fact that the virtual-reality component deepened the empathy I felt toward those people. Even though the video was only a few minutes long, it was a powerful, immersive experience. I felt as if I was in the scene, which is something that simply cannot be replicated by reading an article or watching a two-dimensional recording.

You can take many everyday experiences and make them better with virtual reality. Most people, if they're lucky, watch television at home on a fifty-inch screen. With virtual reality, you could feel as if you're watching on a one-hundred-inch screen, having an IMAX theater experience, or watching a drive-in movie. You could travel anywhere through virtual reality, which may serve as a meaningful replacement when physical travel is unfeasible. For example, if you have a whole class of students who are studying the Battle of Gettysburg, one option is to rent a bus and drive them all the way out to a reenactment. The other option is to place 360-degree cameras on the field and have the students experience the event through virtual reality. If a student missed an explanation, he or she could even rewind the video.

Virtual reality could help minimize the amount of travel people have to do for business as well. Imagine that, instead of going to a physical conference, you could attend a conference in a virtual-reality setting. You wouldn't get jet lag or have to be away from your family for a week, yet you'd have the same human connections that make attending a conference so valuable. Virtual travel could save people time and money, and it would also benefit the environment by cutting down on carbon emissions. In these ways, virtual reality could make the world a little better for people.

Going forward, we have to be careful about how we use the ability to evoke strong emotions in people. As inspiring as an experience could be, one could also be equally horrifying. Imagine if you felt fully immersed in a scene and saw someone get shot, or if someone appeared to attack you. The risk is that a person could be traumatized by a simulated experience and react in a similar way as if they had actually gone through the events. We could see people developing post-traumatic stress disorder if they watch the wrong thing in virtual reality. At the same time, the technology could be an effective tool for treating post-traumatic stress disorder and other conditions, so in that way it's a double-edged sword.

None of the progress being made in virtual and augmented reality would be possible without the advancements in mobile technology. The toolset created by mobile—low-cost hardware, accurate sensors, and low-persistence screens—is what allowed virtual reality to springboard out of our imaginations and into the palms of our hands. Soon, the technology will be affordable and available to everyone.

In the near future, we'll see virtual reality disrupting the status quo in education, health care, real estate, entertainment, and business. Perhaps it will affect industries in ways we can't even foresee, but one prediction that I believe we can confidently make is that virtual reality's impact is going to be profound. We're entering a new age of creativity where we can create environments and experiences in any way we see fit. Every dreamer will be an architect, and with fully immersive virtual reality just over the horizon, we can all look forward to exploring a limitless new world.

WILL TECHNOLOGY LOVE ME BACK?

Gary Clayton

GARY CLAYTON is the former Chief Creative Officer for Nuance Communications, where he held a unique role that sat at the intersection of strategy, innovation and design. Spanning the healthcare, consumer and enterprise markets, Gary's team interpreted technology trends through a design lens to create next-generation user experiences. While at Nuance, Gary was instrumental in creating Nina, Wintermute, DragonTV, Dragon Reader, Dragon Go (Time Magazine Best Apps of the Year) and Dragon Dictation (Apple's App Hall of Fame). Prior to joining Nuance, Gary was the Vice President of Speech Strategy for Yahoo! Where he invented unconstricted speech-enabled web search. Prior to Yahoo!, Gary was the Chief Creative Officer for Tellme Networks where he authored the company's value proposition as it moved into the enterprise. Prior to Tellme, Gary was the founder of Clayton Multimedia where he specialized in media production including film, albums, toys, TV and all manner of media. Gary holds fifteen patents and applications, is a Grammy Award nominee, and his innovations have been acclaimed by a number of associations and journals, including the Wall Street Journal Innovation Awards. He is a voting member of the Grammys and the Emmys.

We've seen it in science fiction for years: devices using artificial intelligence to act as caretakers, personal assistants, and companions. They're able to converse fluently with us and answer our questions intelligently. They get to know and understand our individual needs. They become part of the family. With each advancement in artificial intelligence and speech recognition, we're getting closer to making that dream a reality—but what will that relationship with technology look like?

We have to start at the beginning of the relationship and think about that first interaction. There is one question that people consistently ask after speaking with a device: "How well does the system understand me?"

Does it understand the words we use? Is it able to capture our meaning? The factor that everybody is concentrating on is the system's capacity to understand human input. It's a critical element to explore, but as someone with an extensive background in speech recognition, I'm also interested in looking at the other side of the equation. I want to ask, "What is the system saying back to me?"

If you've interacted with Apple's Siri before, you likely recognized the system's voice as soon as it gave you an answer to your question. Maybe not that exact voice, but you recognized it as an artificial intelligence. When we look at various artificial intelligence systems from major technology companies—Apple's Siri, Microsoft's Cortana, Google Now, Dragon Mobile Assistant—it's easy to hear that they have certain traits in common. They're all female voices of roughly the same age and likely based on a professional voice artist. They sound remarkably similar to one another, and therefore our relationship with them is likely to be similar, regardless of which artificial intelligence we engage. You may or may not have realized it at the time, but as soon as you interacted with that artificial intelligence system, you entered into an implicit relationship with the technology.

When we look at that relationship, we should ask, "Will technology love me back?" So often we develop attachments to our devices, but will we program them to reciprocate? As people, we tend to anthropomorphize our devices, and that's where much of this emotional attachment originates. Technology may not have allowed us to interact at this level until the twentieth century, but the desire to speak to technology is not a new one.

Humanity has a long history, going back thousands of years, of efforts to make technology speak to us. In the eleventh century, we had brazen heads, which through the use of incantations were expected to come to life and talk to their creators. Over the intervening centuries, we created mechanical devices, bellows-driven machines, and other inventions intended to replicate speech, but it wasn't until the mid-twentieth century that we finally made real progress.

As we developed this new technology, we had to decide, to what end? Much of the focus of our modern efforts to create better artificial speech has been on assistive technologies. One of the most famous examples of this is Stephen Hawking and his voice, Perfect Paul. There's been ELIZA at the Massachusetts Institute of Technology, Apple's Talking Moose, and Microsoft's Clippy. All of these personas were designed around the paradigm of the personal assistant. We've seen the progression of technology go from personal assistants to the Internet of Things, in which all of our devices are connected to the Internet and often connecting with one another, and finally to robotics. These three trends—personal assistants, interconnected devices, and robotics—are part of the same larger trend. They're all dependent on forming a relationship with the user.

We choose to design machines in a way that will foster that relationship, and this choice can be seen in projects from top robotics researchers. For example, Boston Dynamics created a robot dog and released a video of the team testing the robot's stability and balance. Someone kicks the robot and, after stumbling to the side, the robot is able to right itself. It's an amazing demonstration of dynamic stability, but I found that my first, immediate reaction upon seeing the video was a twinge of outrage. I had to pause for a moment and ask myself, "What is making me react this way? What is causing this emotion?" It was the fact that I had anthropomorphized the device. Instead of seeing a man kicking a robot, my brain said, "That's a dog. Why did he kick it?"

Man's best friend is a recurring form that we continue to see in robotics. Sony developed a robotic dog called Aibo that struck a particularly strong emotional chord with many consumers. Multitudes of documented anecdotes describe people becoming attached to their robotic pets—for example, stories of individuals who wouldn't leave the house before putting Aibo to bed. They formed relationships with technology—real relationships. However, our relationship with technology is not always a positive one.

There's a dialogue happening around artificial intelligence and the possibility that we might let the genie out of the bottle. These are fears that Elon Musk and Stephen Hawking have publicly expressed, but I wonder how much of that fear is based in fiction. If you think back to the classic science-fiction film *2001: A Space Odyssey*, you may recall the iconic scene in which HAL, the spaceship's artificial intelligence system, rebels against his creators. The relationship between HAL and the astronaut, Dave, changes throughout the movie, with HAL acting as a genuine personal assistant in the beginning and eventually becoming an ominous force. The change is largely demonstrated by what HAL chooses to say and the way in which he says it. We see that the voice plays a key role in the power shift that takes place. Just as HAL's tone and persona dictated how the audience perceived him, so also do the voices we hear in the devices around us affect our perception of technology. The output, the way technology speaks to us, is a critical point of consideration when designing new systems.

As we advance into mainstream robotics, the relationship between humans and technology is going to change. We already have personal connections with our phones, but it's just that—personal. We typically don't share our phones, and our relationship with the device fits the one-on-one personal assistant model. However, with robots it will be different. There will be more of a social component because the robot will need to interact with multiple people. If a household owns a robot, it may even be thought of more as a member of the family or a pet than an appliance. Everyone in the family is going to interact with it, and each will have a different relationship with the device. The artificial intelligence is going to have to be more flexible, more adaptable, and more intelligent. It will have to learn how to speak with each person, and it will most likely talk to the adult members of the household quite differently than the children. The introduction of the social robot creates a whole new level of personalized interactions, and it's up to us to decide how those different voices will sound.

In making these decisions, technologists at leading companies in the industry, such as Apple, Google, and Microsoft, are talking about what is required to create a personal assistant. They're discussing potential applications, the different scenarios in which the device may be used, and what would be required to sell it. Collectively, we've spent so much time focusing on artificial intelligence as an assistant, but what about the personal part? What if the killer application turns out to be companionship?

Imagine a robot sitting on the table next to every bed in the pediatric ward of Boston General. What happens late at night when that child wakes up scared, turns to the robot, and says, "Tell me a story"? What is the persona that responds? The relationships that form between people and technology are not fake. They're real relationships. We need to think about what that relationship will look like and how that artificial intelligence will sound. How will it react? To answer these questions, we'll need to continue developing natural language, carefully plan the way we use speech, and keep the relationship at the forefront.

We'll have to decide whether technology will love us back.

SPECIAL OFFER FOR READERS

Mobile Industry Statistics

The mobile industry is growing rapidly and ecosystem statistics would already be out of date by the time this book went to press. In light of this, we have curated the best mobile industry statistics exclusively for you, the readers of Strictly Mobile. Head to the Strictly Mobile stats page at **www.strictlymobilestats.com** and use the code: **Strictly16** for access and sign up to receive alerts and updates.

Join the Discussion

Visit us online at **www.strictlymobilebook.com** for updates, to meet the authors and join the discussion on the future of mobile.

Follow us

@strictly_mobile

Made in the USA
San Bernardino, CA
03 May 2016